Standardization
Essentials

Standardization Essentials
Principles and Practice

Steven M. Spivak
*The University of Maryland
College Park, Maryland*

F. Cecil Brenner
*Tire Technology and Standards Consultant
Silver Spring, Maryland*

MARCEL DEKKER, INC.   NEW YORK · BASEL

ISBN: 0-8247-8918-0

This book is printed on acid-free paper.

Headquarters
Marcel Dekker, Inc.
270 Madison Avenue, New York, NY 10016
tel: 212-696-9000; fax: 212-685-4540

Eastern Hemisphere Distribution
Marcel Dekker AG
Hutgasse 4, Postfach 812, CH-4001 Basel, Switzerland
tel: 41-61-261-8482; fax: 41-61-261-8896

World Wide Web
http://www.dekker.com

The publisher offers discounts on this book when ordered in bulk quantities. For more information, write to Special Sales/Professional Marketing at the headquarters address above.

Current printing (last digit):
10 9 8 7 6 5 4 3 2 1

PRINTED IN THE UNITED STATES OF AMERICA

In memory of
Dr. F. Cecil Brenner, 1918–1998

Foreword

Standardization is indispensable to life in this century. It is virtually as indispensable as the air we breathe. And, like the air we breathe, it is invisible to all except its technicians.

Standardization impacts life dramatically. It means progress. It shapes technology and defines the terms of commerce. It sustains our environment and makes us safe. It has given us the wherewithal to walk on the moon. And yet it is invisible.

In spite of its importance, no great body of literature is devoted to the study of standardization. This comprehensive and scholarly book is a benchmark publication that brings standardization into public visibility by presenting the subject from its basic principles to its impact on the world of trade. Standardization is now recognized, as confirmed in a European paper, as something far too important to be left to the sole purview of technicians. It is everybody's business. This book, therefore, could not come at a better time.

In this unique publication, the authors help us to understand standardization on all its levels. They unravel its structures, language, and processes. Through the experiences of guest authors, the principles of standardization are set in real-life situations, and we are led to the understanding that standardization is not the ward of science alone, nor can it be defined by its structures. First-hand accounts demonstrate that standardization is also the art of agreement and often the art of politics. I know of no other publication that offers this practical blend of theory and practice. For this reason alone, it should be valued as a reference and used as a guide by seasoned standards practitioners as well as those new to the field.

We owe a debt of gratitude to the authors. The late Dr. Cecil Brenner and his pioneering work in the standardization of tires were well known to me. He was a friend to me and to the organization of which I am so proud, the American Society for Testing and Materials (ASTM). He wrote much of what is written about ASTM in this book. He will be greatly missed.

Dr. Steven Spivak is a respected colleague, and also a friend. He is a rare breed, an academician who understands standardization deeply and fully, and who has labored tirelessly to dispel the clouds of ignorance that surround it. Thank you, Steve, for carrying on, and bringing this fine work to fruition.

James A. Thomas
President, ASTM
West Conshohocken, Pennsylvania

Preface

Standardization Essentials is both a primer and a review of standards and standardization. For a field as vital and pervasive as standardization, there are few books that cover the basic principles and practice, as well as share guidance and practical experiences on the actual workings of standards-developing organizations and standardization. The book begins from first principles of standards and standardization and gives definitions for common terms and descriptions used in this field. It explains how standards could be misused in restraint of trade and how that misuse can be prevented by standards-writing organizations and their due process provisions and globally through appropriate international representation. The book details how international trade is fostered by judicious use of standards and where standards might become impediments to regional or global trade. It also covers the essential role that standards play in strategic standardization management, in purchasing and contractual agreements, in quality assurance, and in many other areas.

Included are the experiences of long-time practitioners in the field and a ready reference guide to the major "players" in the field of standardization. These include the national standards bodies or coordinators and standards-developing organizations (SDOs)—traditional and nontraditional. Although the major focus is the United States and Canada, with some reference to Mexico and the tripartite North American Free Trade Agreement (NAFTA), there is also extensive coverage of global standardization. A glossary defines many abbreviations and acronyms used in standardization and identifies or explains the numerous array of abbreviations for standards organizations, often seen but seldom understood.

The audience we had in mind includes engineers and managers, procurement personnel, strategic standardization and trade promotion specialists, quality control engineers, quality management professionals, and those embarking on a process of standards development in any field of endeavor. Other interested readers would include information technology and technical information specialists, librarians, and faculty and students in business and management, engineering and engineering technology, human ecology, and the information sciences. The book is intended for all those who discover, need to know, and seek greater understanding of the essentials and elements of the discipline of standardization.

The book should be particularly helpful in the activities of ISO in international standardization, specifically ISO 9000, Quality Management Standards, ISO 14000, Environmental Management Standards (EMS), and strategic standardization and international trade. These sections provide an overview and background to the many business and industry people who have heard of ISO 9000—and may even be intimately involved with its standards—but do not know the source or purpose of ISO 9000. In the international arena, the World Trade Organization (WTO) is touched on, as is its predecessor, the General Agreement on Tariffs and Trade (GATT). Their "Standards Code" with general principles on standardization, and now trade in services and related service standards, become increasingly important to both governmental and nongovernmental organizations.

Part I, "Fundamentals of Standardization," covers the essential information needed by those entering the field. After relating historical anecdotes from earliest times to recent years, we discuss the semantics of standards, including definitions, terms, and descriptions of various practices. The purposes, scope, and principles of standardization are followed by a description of how various kinds of standards are developed by different standards-writing organizations. These include both voluntary (or nongovernmental) standards and mandatory (or governmental) standards and regulations. Standards often lead both lives, starting out as voluntary but later being adopted or referenced into local, state or provincial, or federal regulation.

Part II, "Elements of Global Standardization," commences with a summary discussion of the elements of national, regional, and international standardization, the last two of which are also referred to as "global standardization." Several national standards systems and the components of the voluntary international system are presented. These are followed by a discussion of the competition surrounding the role of national versus international standards.

A brief overview of ISO 9000 and ISO 14000 management system standards precedes an elaboration of implementing ISO 9000 quality management system standards in industry. There follows a new summary and perspective on the U.S. standards system in relation to its European and international competitors and partners. These sections should be helpful to business and quality management professionals who find themselves having to consider or implement ISO

9000 or ANSI/ASQ 9000 quality management standards or ISO 14000 environmental management standards. Because such situations are common in business,
industry, government, and academia, our intent in this book is to supply the necessary background and practical information. In this context the book serves as a
companion to others published by Marcel Dekker, Inc., on the topic of ISO 9000
and management system standards—both quality management systems (QMS)
and environmental management systems (EMS). The corollary of developing regional or internationally harmonized standards for occupational safety and health
(OS&H) management system standards, however, remain controversial at the
time of this writing.

We continue with the issue of standards as builders or barriers to trade,
specifically the use of international standards to leverage world markets. We present a case study of conformity assessment related to international technical trade
barriers and a discussion of safety standards and product certification in the global
marketplace. We conclude Part II with a focus on consumer goods and consumer
standardization, for products such as automobile tires, toys, textiles, and furnish-
ings, product safety, and for consumer services. The use of standards to facilitate
movement of consumer goods worldwide and companion issues of product safety
are highlighted.

Part III, "Reminiscences of Experienced Standards Makers," is a series
of vignettes by invited authors experienced in the development, policy, and practice of standardization. In relating their personal experiences, they explicate some
of the social, societal, and political aspects of the field. These add a practical focus
to standards activities that might otherwise appear to be devoid of a subjective
component. The contributors present information about alternative or consortia
development of standards; the uses, costs, and benefits of effective standardization
programs; and even limitations and some abuses of standards and standardization.

Part IV, "Standards Organizations," is an alphabetical listing of many
common standards organizations, their acronyms and names, headquarters locations, and brief descriptions of what they do. It will serve as a ready reference
for readers using the World Wide Web or contacting SDOs and standards bodies
directly to obtain further details, catalogs of standards, or other documents. Coverage includes organizations in the United States, Canada, and Mexico, plus regional and international activities. We note that while many professional and
technical societies, associations, and organizations are included, there are perhaps
a thousand more just as important in their own right. There is no suggestion that
they are less significant than those included here.

Readers with no previous experience in the subject matter may find it beneficial to first read one or more chapters of interest to them in Parts II or III before
taking on the more technical subjects in Part I. Since Parts II and III comprise
essays detailing the experiences of "practiced hands" in the field of standardization, these chapters can tell and teach much.

In acknowledgment, we express our gratitude to Eudice Segal and Margaret "Peg" Brenner for the many editorial changes they suggested, and to Thomas Powers for his yeoman's task of proofreading the entire manuscript. Sanjeev Gandhi and Steven Wolin offered computer skills to assist with document and figure preparation and related expertise, for which we thank them accordingly. We are also grateful to the American National Standards Institute (ANSI; New York, NY, and Washington, DC), the American Society for Testing and Materials (ASTM; West Conshohocken, PA), the Standards Engineering Society (SES; Miami, FL), and the International Organization for Standardization (ISO; Geneva, Switzerland) for permission to use their copyrighted material, cartoons, or excerpts from articles for this book.

We gained invaluable standards experience and professional friendships from ASTM, the National Fire Protection Association (NFPA), and others whose consensus standards development processes are examples of how to do it right. We also thank the contributing authors, colleagues plus several current and past members of the ANSI Board of Directors, for their willingness to share reminiscences and experiences with our readers.

The consumer standardizers among ASTM and ISO, especially the latter's dynamic Consumer Policy Committee (ISO/COPOLCO), deserve our gratitude and appreciation for many accomplishments in fostering the standardization of performance, health, safety, labeling, and quality of products and services worldwide. Beatrice Frey and the expert ISO staff have ably managed the secretariat for COPOLCO (and ISO's Central Secretariat) for many years and are respected colleagues in this mutual journey.

In particular, we acknowledge Giles Allen of ISO (also one of the expert writers for the ISO Bulletin) as well as cartoonists, including Pascal Krieger, for generously allowing use of their lively and skillful cartoons. Standardization is never dull—it is often demanding or challenging and, at times, exasperating. But for the neophyte to standardization, certain books may at first glance engender all these adjectives, thus our liberal use of the ISO Bulletin's witty cartooning to enliven and focus the importance and value of these subjects.

Lastly, a note of remembrance to the late Dr. Lal C. Verman, whose tome *Standardization: A New Discipline* is an inspiration. One of the authors (S.M.S.) strove to see Verman's seminal book republished in North America, India, and Asia, but without success. Readers seeking detailed discussion of standards information and standards education as well as standards used in the fields of information technology and archival and information standards can consult Spivak and Winsell's *A Sourcebook of Standards Information: Education, Access and Development* (New York: G. K. Hall, Macmillan, 1991, ISBN: 0-8161-1948-1).

Steven M. Spivak
F. Cecil Brenner

Contents

Contributors

Balbir Bhagowalia   Former United Nations (UNIDO) Consultant, Sterling, Virginia

F. Cecil Brenner[†]   Tire Technology and Standards Consultant, Silver Spring, Maryland

Helen Delaney   Delaney Consulting, Bethesda, Maryland

Carl Cargill   Standards Management, Sun Microsystems, Menlo Park, California

Norm Hagan   Norm Hagan & Associates, Westport, Ontario, Canada

David E. Jones   Quality Assurance, Robinson Brothers Limited, West Bromwich, England

Frank Kitzantides   National Electrical Manufacturers Association (NEMA), Arlington, Virginia

Stephen C. Lowell   Standardization Program, U.S. Department of Defense (DoD), Fort Belvoir, Virginia

---

[†]Deceased.

Donald R. Mackay   International Standards, Air-Conditioning and Refrigeration Institute (ARI), Arlington, Virginia

Stephen P. Oksala   Corporate Standards Management, Unisys Corporation, Malvern, Pennsylvania

C. Ronald Simpkins   Consultant, Kemblesville, Pennsylvania

Nancy Harvey Steorts   Nancy Harvey Steorts International, McLean, Virginia

Keith B. Termaat   Cross Platform Closures, Ford Motor Company, Dearborn, Michigan

James A. Thomas   American Society for Testing and Materials (ASTM), West Conshohocken, Pennsylvania

Robert B. Toth   R. B. Toth Associates, Alexandria, Virginia

Peter S. Unger   American Association for Laboratory Accreditation (A2LA), Frederick, Maryland

Robert A. Williams   Standards and Research, Underwriters Laboratories, Northbrook, Illinois

1

Introduction

This introduction and overview to the book also includes portions of "Quality, Performance and Standards for Textiles," written by one of the authors, Steven Spivak, as the opening chapter to the special thematic issue on Quality for the *Journal of the Textile Institute*, 1994. Dr. Spivak also served as the guest editor for that special issue.

STANDARDS DEFINED

Standards, quality, and *performance*—what do the terms mean and how do they interrelate? In the simplest sense, a *standard* is an agreed-upon way of doing something. Slightly more elaborate, a *standard* denotes a uniform set of measures, agreements, conditions, or specifications between parties; the latter may be buyer-seller, manufacturer-user, government-industry or government-governed, retailer-manufacturer-consumer, or any other parties. Commerce and trade are built on a foundation of rational standards.

A complete set of definitions for *standardization* and *standard*, types of standards, terms for *performance, testing and certification, laboratory accreditation*, and *quality control* can be found in *A Sourcebook on Standards Information: Education, Access and Development* (1). There are numerous versions of the common terms and definitions used in conjunction with standardization, quality, and performance of any product, service, or system for that matter. From the International Organization for Standardization (ISO) the *process of standardization* is defined as "standardization: activity of establishing with regard to actual or potential problems, provisions for common and repeated use, aimed at the achievement of the optimum degree of order in a given context. Note: In particular, the activity consists of the processes of formulating, issuing and implementing standards.'' Likewise, the ISO definition of standard: ''Document, established by consensus and approved by a recognized body, that provides for common and repeated use, rules, guidelines or characteristics for activities or their results, aimed at the achievement of the optimum degree of order in a given context.''

TYPES OF STANDARDS

Standards are not monolithic, nor are they of the same general type, acceptance, or function. Indeed, much of the current work of standardizers is aimed at somehow harmonizing or rationalizing the diversity among national, regional, and international standards. Let's look at these issues and see why. Standards come in many forms. They may be physical items or things, particularly standard units of weights and measures, or they are written documents, published norms. It is the latter with which we will be primarily concerned, in other words, standards as reference documents, published or increasingly available in electronic format whether on CD-ROM or online from the publisher or distributor.

Particular applications of standards may be, but are not limited to:

1. Physical standards or units of measure;
2. Terms, definitions, classes, grades, ratings, or symbols;
3. Test methods, recommended practices, guides, and other application to products and processes;
4. Standards for systems and services, in particular quality standardization and related aspects of management system standards for quality and the environment; and
5. Standards for health, safety, consumers, and the environment.

In this book the authors' foci are on both traditional and more-recent evolving issues of standards, globalization and trade, national versus regional versus international standards, quality and total quality management (TQM), process quality control and standards, health and safety standards, consumer product and service standards, and much more. Standards for quality and quality systems are a good place to start for several reasons. First, they are the fastest moving issue in the 1990s and the first years of the new millennium. Second, quality standards are now developing rapidly throughout the full spectrum of business, industry, and government. Quality management standards have become the ''norm'' by which many, unfamiliar with standards organizations and standards developers, have heard of what it is we do as standardizers. Let us explain, beginning with quality standards and quality systems registration and ISO 9000 (2).

ISO 9000 quality-management system standards are fast becoming, if not already, the *sine qua non* of quality-conscious industry and standards professionals. ISO 14001 environmental-management system standards are going to have a similar, dramatic impact on business, industry, associations, and government. Even if one does not quite know who is the ISO or mislabels their moniker as International Standards Organization, many professionals have still heard of ISO 9000 or ISO 14000 series standards, or their national-equivalent management systems standards. Total quality management principles, and practices of continuous improvement including quality systems registration, are becoming the "norm" or benchmark against which other entities are measuring themselves. Of the more than 10,000 ISO voluntary international standards, and another 2,500 standards of its compatriot organization the International Electrotechnical Commission (IEC), none have become as popular or widely accepted as are the ISO 9000 set of international quality-management system standards. This is the subject of two chapters in this book.

LEVELS OR HIERARCHY OF STANDARDS

This book and its readership are international in scope, and so are the standards of the ISO and IEC. But this international focus highlights another critical trend in today's global marketplace, that of the interplay and competition between the various levels (or hierarchy) of standardization. Let's explore these levels, and fit them into a three-dimensional standardization space first elaborated by Verman (3) in his seminal work, now out of print, *Standardization: A New Discipline*. At the highest (or, if you prefer, broadest) level of applicability are the international standards. Voluntary international standards of prevalence and interest to managers, engineers, and educators would be those standards developed and published by the ISO. These include numerous standards, guides, and documents emanating from ISO's 200 or so technical committees (TCs) or from other related subcommittees such as air quality or water quality; or quality-management system standards (2) such as the ISO 9000 and ISO 10000 series developed by ISO TC 176; or ISO 14000 series standards for environmental management systems (EMS) produced by ISO TC 207. Industry-specific variations of the management systems abound with QS9000 (automotive), AS9000 (aerospace), SA8000 (social accountability), and many more.

At the regional level, standards are developed that are germane to a specific geographic region. In the European Economic Union (EU, EC, EEC) and neighboring European Free Trade Association (EFTA), it is European Norms (ENs) that are their growing body of regional standards. Developed by CEN and CENELEC, the European committees for standardization (normalization) and electrotechnical standardization respectively, there are, for example, the quality system standards series EN 29000 and EN 45000, essentially identical versions to the ISO 9000 and 10000 series of standards.

At the national level, almost every nation of the world has its own body of published standards. In some cases, the respective national standard is as well or better known within its own borders than are the international standards. Again using quality system standards (or other standards for that matter) as examples, in the United Kingdom it is British Standards (BS) of the British Standards Institution (BSI) that prevail. Thus, BS 5750 quality-management system standards were both the forerunners of ISO 9000 and are still the well-known designation for equivalent, British national quality standards. Indeed, most nations now have published national standards that are identical or essentially the same as the ISO 9000 series, and are now following suit with the ISO 14000 series.

In the United States, the ISO 9000 standards are equivalent to the national ANSI/ASQC 9000 series quality standards, published by the American Society for Quality (ASQ) and adopted by the American National Standards Institute (ANSI) as identical national standards. In Canada, roughly similar national standards to the ISO 9000 exist as Canadian national standards, as do identical versions in France (Association Francaise de Normalisation, AFNOR), Germany (Deutsche Insitut für Normung, DIN), India (Bureau of Indian Standards, BIS), Denmark (Dansk Standard, DS), and throughout the world. In addition to those enumerated here, hundreds of other standards developers publish product or system-specific standards and related documents that are also used both nationally and internationally.

One must also recognize the less seen, but equally important, body of company or corporate standards, comprising the largest total number and usage of standards. But these are generally unpublished and unavailable, except to those doing business and having the need to know of a specific company standard. We will, therefore, restrict our discussion to the better-known and broadly available published standards.

STANDARDIZATION SPACE: AN ORGANIZED APPROACH TO STANDARDS

One question that arises among those with organized or "standardized" thought processes is how are the diverse standards systems related, or how might they be organized in a coherent way? What is the relationship between different types of standards, their applicability to many aspects that are standardized, and the level or hierarchy at which these standards operate or are used? Although described in greater detail in Chapter 4, it is worth introducing the elegant scheme first presented by the late Dr. Lal Verman (2). He devised and illustrated a schema for standardization in three dimensions or 3-space, which has since been elaborated upon in greater detail by others, and it remains valuable.

In the vertical (z) direction is the level of standardization work discussed previously, be they company standards, trade association, professional or technical society, local, state and province, or national standards authorities; or regional and international standards. This hierarchical approach has since been expanded upon to show the important distinction between voluntary (nontreaty) and mandatory-regulatory (or treaty) standardization developments and practices. Verman shows this hierarchy as a pyramid with a broad base and large number of company standards, rising up the pyramid through variety reduction and lesser total numbers of standards but at higher levels of applicability or influence. International standards (e.g., ISO, IEC) appear at the pinnacle of the pyramid, but there are many who would argue that other national or society standards have equivalent international applicability and importance. Still, it is a useful point of departure for better understanding and debating the global standards system in a somewhat organized manner. In the x axis are the subjects of specific standardization work, such as textiles or machinery, air and water quality, or product performance and quality standards. The y axis is the aspect or type of standardization work, describing the nature of the standards rather than the specific product(s) or service(s) to which they apply. Under aspects or types of standards would be terminology and symbols, test methods, certification, laboratory accreditation, codes and regulations, guidelines or industry practices, safety and health standards, quality system standards, and many more. The z axis or vertical planes represent the various levels of national, regional, and international standardization. One needs to add company standards, and local province or state authority standards where applicable. In addition, there is the distinct body of both voluntary and mandatory standards. All standardization work can be represented or characterized within this three-dimensional standardization space. It is a useful illustration and representation, as will be seen later in the book.

VOLUNTARY OR REGULATORY USE OF STANDARDS

In addition to the development and publication of codes or standards, be they for specification, quality aspects, or performance, health, and safety, their ultimate usage in the marketplace may remain voluntary or become mandatory. This is important since mandatory standards are often cited as reasons for technical trade barriers. An excellent exposition and analysis of these themes is available, the subject of a major study by the committee on consumer policy of the Organization for Economic Cooperation and Development (OECD) (4).

Standards referred to or used by specialists often begin life as voluntary, having been developed and published by voluntary standards organizations such as the ISO, BSI, DIN, ASTM, IEEE, and others. At the international level, such voluntary standards are recognized as the domain of nontreaty standards developers. Many standards are used voluntarily, but others may also undergo a transition in which they become adopted or referenced into mandatory regulations. Further, other standards are written and published by national or local governing authorities, with the expressed intent of becoming mandatory or regulatory in function and use. When this happens at the international level, such mandatory standards are the domain of so-called treaty organizations and their signatories.

One premier example of a treaty body involving standards and standardization is the World Trade Organization (WTO), having superseded the General Agreement on Tariffs and Trade (GATT). Of specific interest to standards personnel are companion or side agreements such as the Agreement on Technical Barriers to Trade, known as the *standards code*. It is here whereby some of our periodic product standards–related *trade wars* are fought over and negotiated. Improving trade in products and in the burgeoning global business of services through recognition and use of international standards and harmonized regional standards, has been the subject of many articles. Examples include those of a standards-and-trade theme issue in *Standardization News of ASTM* (5) and many others where the underlying issues and importance of strategic standardization management have remained the same.

PERFORMANCE, SAFETY, AND QUALITY

Building on this brief introduction to standardization, this book also focuses on the issues of total quality management; the experiences with quality-management system standards and quality registration/certification in North America and in the United Kingdom; and process and product-quality improvement. This theme is continued in Chapter 18 as it covers safety, quality, and performance aspects directed to meet consumer needs. In this chapter are included major aspects of consumer product standards, and health and safety aspects.

REFERENCES

1. SM Spivak and KA Winsell, eds. A Sourcebook on Standards Information: Education, Access and Development, G. K. Hall Reference. New York: Macmillan, 1992, pp. 313–333.
2. ISO 9000–9004 International Quality Management System Standards, 2nd ed. and A Compendium of Quality Standards, Geneva, Switzerland: International Organization for Standardization, 1994. The compendium is also available from your national standards body.
3. L Verman, Standardization: A New Discipline, Hamden, Connecticut: Archon Books, 1973. Also published in India and out of print in both North America and Asia.
4. Consumers, Product Safety Standards and International Trade, Committee on Consumer Policy, Paris, Organization for Economic Cooperation and Development, OECD/DAFFE/CCP, 1986. Also available at the OECD bookstore, Washington, D.C.
5. JG O'Grady, CM Ludolph, JR Woods, H Davis, PJ Apostolakis, et al., ASTM Standardization News, 2:24–42, 1990.

2

Through the Ages with Standards

In early days, people's need to measure things marked the beginning of the development of standards. For example, when man left the fields and founded cities, he needed a measure of length to build temples and palaces and some measure of volume to tithe grain to the priests and kings. The story of setting standards then sprang from the simple need to assign practical and fair value to space and objects. Of course, the earliest trade was carried out by barter and did not require measures for quantity, but by 3000 BC weighing became a factor in trade. Simple weights and measures as national references were deposited in the principal temples (1). Around 2500 BC, small balances were devised for measuring gold dust and by 1350 BC these balances were commonly used in trade.

In the earliest societies, the dimensional standards were taken from the human anatomy, frequently the king's body. The "cubit" was the distance from the elbow to the wrist and has been found to vary from 0.641 meters for the Roman cubit to 0.444 meters for the Palestinian cubit among at least eight different cubits (2). Two common shorter lengths were the "span," the distance from the tip of the thumb to the tip of the little finger with fingers extended (modern standard is 9 in., 23 cm) and the "hand" (4 in., 10 cm). This latter measurement is still used to specify a horse's height, in other words, so many hands tall.

We also find the cubit as a unit of measurement mentioned in the Old Testament. In Exodus 25:10, the Lord instructed Moses to make "an ark of acacia wood, two and a half cubits long, a cubit and a half wide, and a cubit and a half high." The Ark incidently was the housing in which the Torah (first five books of Moses) was carried across the Egyptian desert (3).

The royal practice of using the king's body as a source of standard dimensions continued through the middle ages and at least into the early Renaissance.

We see in 1120, Henry I of England decreed the ell, the traditional cloth measure, as the distance from his left shoulder to the tip of his extended left hand. The ell eventually became the basis for the yard (0.91 m) (4).

Weighing precious metals or gems was done on a wooden balance arm suspended by leather thongs or yarn at the midpoint with containers suspended from the arm ends. The weights were seeds or kernels of grain; the carob seed was the precursor of the carat, in which modern gems are measured. In the Middle Ages, Henry III, 1216–1272, devised the troy system for weighing precious metals based on the pennyweight, which was equal to thirty-two grains of wheat. The name came from weights used at the Troyes Fair, which had its own weights and coins (5).

Verman (6) suggests that even prehistoric man had standards. He cites as evidence the similarity of stone tools. Within some cultures the similarity is striking. For example, every Clovis point has a unique scalloped edge. However, we do not believe that these arose from the application of standards in the modern sense. They probably represent the influence of a dominant or gifted teacher who instructed novices in "proper" technique or possibly as an influence of tradition. We say this because there appears no rational reason for the adoption of such standards. The influence of a dominant teacher is known from our experience with laboratories engaged in testing materials in which technicians manually handle materials. Frequently, this handling cannot be described so that all technicians carry out the handling with the same result. However, technicians trained by the same supervisor do reproduce the supervisor's results. We find large constant differences between laboratories, each of which is self-consistent. That is, there are only small differences in results among technicians in the same laboratory. One method to bring the laboratories into agreement with each other is to train all the supervisors in one industry laboratory (7).

EARLY RECOGNITION—SAFEGUARDS AGAINST CHEATS

Honest standards were early recognized as important to society and were given biblical authority. In the third book of Moses (Lv 19:35–36), the Lord commands, "You shall not falsify measures of length, weight or capacity. You shall have an honest balance, honest weights, an honest *ephah*, and an honest *hin*" (*ephah* = a little more than a bushel; *hin* = a gallon and a half) (3). And in Proverbs 11:1, we find "A false balance is abhorrent to Yahweh, a just weight is pleasing to him" (8).

Not only in biblical text, but also in Indian literature there were cautions against cheating. Verman (6) points out "In the Sanskrit scripture of about the same period as Western biblical times one finds similar references. For example, Manu (a "divinely inspired law giver" (9)) in his Manusmriti (about 400 BC gives a table of 13 units of weights, and their inter-relationships. He goes on to legislate that 'the king should inspect the weights and measures and have them stamped every six months and punish offenders and cheats.' ''

ARCHIMEDES' PRINCIPLE

History confirms that the biblical concern for cheats was justified. Hiero II, the Greek ruler of Sicily in 215 BC was suspicious that a gold crown that he purchased was diluted with silver, and he called on Archimedes to find out if he had been cheated. There was no known method to determine if the gold was diluted but, as is well known, Archimedes, while bathing, suddenly realized that equal weights of gold and silver, being of different density, would displace different volumes of water. That is, equal weights would occupy different volumes. This enabled him to test the crown and answer the king's question. His finding is not reported (9).

ENGLISH EXPERIENCES

King John and the Magna Carta

On June 15, 1215, King John was forced to sign the Magna Carta at Runnymeade by the English Barons in an attempt to recover some of the rights granted by earlier rulers or established feudal customs long in effect prior to John. John had violated the earlier rules for his personal benefit or gain. Among the some seventy articles in the Magna Carta was Chapter 35, which dealt with "measures." It read "Let there be one measure of wine through our whole realm; and one measure of ale; and one measure of corn, to wit, 'the London quarter' and one width of cloth (whether dyed or russet, or 'halberget'), to wit, two ells within the selvedges; of weights also let it be as of measures."

Prior to 1215 various ordinances (assizes) were issued to regulate the sale of commodities. Assizes were issued to cover beer, wine, and bread and also assizes on weights and measures. Richard the Lionhearted published *Assize of Cloth* (1197), which tried to overcome the inconvenience experienced by traders who met with varying standards as they moved their goods from place to place. Of more importance this assize sought to prevent frauds being perpetuated on buyers under shelter of ambiguous weights and measures; thus the requirement for "one measure" for each of the commodities—ale, corn, etc. Dyed cloths were required to be uniform all across the fabric. Merchants were prohibited from hanging over their windows "cloth whether red or black, or shields so as to deceive the sight of buyers seeking to choose good cloth" (10).

One of the complaints against John was that he did not attempt to enforce the regulations but was content to fine the violators with the fines going to the Royal Treasury! Clearly, these references reflect the concerns for the potential of abuse of weights and measures that had already been established as "standards of the realm."

BAKER'S DOZEN

In sixteenth-century England, the bakers were reputed to be short-weighting their bread. The penalties decreed to eliminate this practice were so severe that the bakers began the practice of giving 13 loaves for each 12 ordered. This is one of the possible origins of the term "Baker's Dozen" (11).

AMERICAN EXPERIENCE

Interchangeable System of Manufacture

Thomas Jefferson, when American Minister to France in 1788, wrote a letter describing the practice of a gunsmith named LeBlanc: "An improvement is made here in the construction of muskets, which may be interesting to Congress to know, should they at anytime propose to procure any. It consists of making every part of them so exactly alike that what belongs to any one, may be used for every other musket in the magazine. . . . Supposing it might be useful to the United States, I went to the workman. He presented me the parts of fifty locks taken to pieces, and arranged in compartments. I put several together myself, taking pieces at hazard as they came to hand, and they fitted in the most perfect manner" (12).

In 1798, the Congress of the United States authorized Vice President Jefferson to contract with Eli Whitney, inventor of the cotton gin, to begin the interchangeable manufacture for muskets. Whitney aimed "to make the same parts of different guns, as the locks, for example, as much like each other as the successive impression of a copper-plate engraving." Whitney apparently did not know of LeBlanc's activity, whose success was probably hampered by the opposition of the skilled gunsmiths of Europe. The United States was perhaps favored for successful introduction of what became known as the American system because of the lack of expert machinists; "a species of skill which is not possessed in this country to any considerable extent" according to Whitney. The establishment of the factory in New Haven, CT took more time than projected, and just when Congress was expressing its frustration, Whitney arrived with finished parts and demonstrated the assembly of ten muskets from randomly selected parts.

THE METRIC SYSTEM

In the last decade of the eighteenth century, the French prompted by Talleyrand began development of a metric system so that standards should be "on a fixed and immutable basis derived from nature." After 160 years of evolution the International System of Units (SI) finally was developed. The history of the evolution is too long and complex for a discussion here; we simply note that the development was in progress.

RAILWAY TRACK GAUGE

Cropper (13), commenting on the difficulty of achieving standardization in practice wrote:

"More turbulent was the establishment of the standard gauge for railroads, which unquestionably came down also from the Romans, but with great and needless travail.

"William Jessop invented metal rails for transportation in about 1795 and laid them the same distance apart as the cartwheel ruts in the English roads of those days. The ruts were probably unchanged from the time of the Romans. In fact, that distance is the same as that later found between wheel-paths in Pompeii and is today our standard railway gauge: 4 ft. 8½ in.!

"However, England's Great Western Railway was built in the early 1800's with a rail gauge of seven ft. In Canada and on some U.S. lines, the approved gauge in the first half of the 19th century was 5 ft. 6 in. By the time of the American Civil War, there were no less than 33 different gauges in this country.

Source: Courtesy of *ASTM Standardization News*, February 1977.

To aid military logistics, the U.S. Congress in 1863 mandated the 4 ft. 8½ in. gauge, based solely on empirical considerations. It took 25 years to complete the conversion of all U.S. lines. In England, the Great Western was not converted to standard gauge until 1892, about the same time the Canadian lines were also converted. Australia still had five different railway gauges, and it was not until 1970 that through trains were able to traverse that continent on rails that were 4 ft. 8½ in. (1.44 m) apart.

"These costly conversions were not by any means confined to the English-speaking world but were almost universally required. This lack of standards had tremendous economic consequences: freight had to be unloaded and reloaded at junction points; rolling stock could only be used for short hauls instead of through runs. Often such large and avoidable costs resulted merely from shortsighted but honest errors in judgment, but there undoubtedly were a great many cases where "maverick" gauges were deliberately chosen for self-serving or preemptive reasons."

SCREW THREADS

In the early nineteenth century one of the greatest obstacles to the easy adoption of the interchangeable system of manufacture was the lack of standard for screw threads. Gilbert (12) observes that "Joseph Whitworth was responsible for bringing about the standardization of screw threads. He collected and compared screws from as many workshops as possible throughout England and in 1841 proposed, in a paper to the Institution of Civil Engineers, the use of a constant angle (55°) between the sides of the threads, and a specification for the number of threads to the inch for the various screw diameters." The Whitworth thread remained the standard in engineering until 1948.

FIRE HYDRANT AND HOSE COUPLINGS

In 1904, a major fire erupted in Baltimore and the local fire department called for assistance from Washington, DC, Philadelphia, and New York. The needed equipment was shipped by express trains to Baltimore to no avail: the hose couplings would not fit the Baltimore hydrants. The problem has been corrected by the efforts of the National Fire Protection Association and others (4).

AMMUNITION

The czarist army suffered a major defeat because all guns of a given caliber did not have exactly the same diameter and a shipment of ammunition was of the

Source: Reproduced from *Through History with Standards*, credit given to the American National Standards Institute (formerly American Standards Association).

right caliber but wrong diameter. During World War I, the War Industries Board assisted in overcoming the shortage of materials and in conserving energy for the war effort. This was largely done by standardization, which reduced the varieties of manufactured items, and ensuring interchangeability of parts. After the war, the gains were made permanent by the adoption of the recommendations in the 1921 Report of the Committee on Elimination of Wastage in Industry. These recommendations were made permanent by U.S. Secretary of Commerce Herbert Hoover. In some industries the variety reduction amounted to 24 to 98 percent (14).

World War II with its greater dependence on advanced technologically developed equipment made adoption of standardization imperative as a means of achieving interchangeability and simplification, and of conserving resources and energy. Lack of standardization on such items as tools, nuts, and bolts, made it necessary to supply U.S. military needs from American sources at great costs in money and time. This drove efforts to improve coordination between standards organizations in the various allied countries that in turn blossomed in the present international efforts in this direction. The International Organization for Standardization (ISO) was formed in the wake of these activities. See Chapter 3 for additional discussion.

AUTOMOTIVE PARTS

The American auto manufacturers have adopted the metric system for all parts except wheel and tire diameters. This enables all manufacturers to buy from and sell parts to all other manufacturers in the world. Similarly, all aircraft parts follow one set of standards throughout the world to facilitate availability. This brief discussion perhaps gives the reader an appreciation of the development and role of standards over the centuries. For anyone who would care to explore some area in detail or browse through many fields we recommend that they look into the five-volume *History of Technology* (Oxford, England: Clarendon Press, 1957), which covers the Stone Age to the twentieth century.

REFERENCES

1. FG Skinner, Weights and Measures In: History of Technology. Vol. 1. Oxford, England: Clarendon Press, 1954, p 774.
2. NIST Handbook 44. Specifications, Tolerances, and Other Technical Requirements for Weighing and Measuring Devices. U.S. Department of Commerce, 1995.
3. The Five Books of Moses. In: The Torah. Philadelphia: The Jewish Publication Society. 1962, pp 143.
4. Through History with Standards. New York: American Standards Association (now ANSI), n.d. In: R. Glie, ed. Speaking of Standards. Boston: Cahners Books, 1972, pp. 37–71.
5. E Weekly. An Etymological Dictionary of Modern English. New York: Dover Publications, 1967.
6. LC Verman. Standardization: A New Discipline. Hamden, CT: Archon Books, 1973 (out of print).
7. FC Brenner. Test Method Standardization—A Dilemma. Clothing and Textile Research Journal 3, no. 1:41–44, 1984–85.
8. The Jerusalem Bible. New York: Doubleday, 1966.
9. The Columbia Encyclopedia. 3rd ed. New York: Columbia University Press.
10. WS McKechnie. Chapter 35. In: Magna Carta. New York: Burt Franklin, 1914.
11. W Morris, M Morris. Dictionary of Word and Phrase Origins. New York: Harper & Row, 1963.
12. KR Gilbert. Machine Tools. In: History of Technology. Vol. 4. Oxford: Clarendon Press, 1958, pp 437–438.
13. WV Cropper. Standards and Standardization. Chemtech (September): 550–559, 1979.
14. LC Verman. Standardization. Hamden, CT: Archon Books, 1973, pp 8–9.

3

Semantics and Terminology

ETYMOLOGY

As we have seen, standards have been in use from the earliest historic times but as far as we know they were not classified as such for many centuries. Our word *standard* comes to us from the common Roman *estend-ere* (Latin *extend-ere* to stretch out). In England, the word first appears in reference to the Battle of the Standard. This was a battle of David I of Scotland against Stephen of England. David had conquered most of Northumberland and Cumberland but was defeated by the English nobles and militia at the Battle of the Standard in 1138; the standard being the flag flown from a ship's mast mounted on a wheeled cart and dragged onto the field of battle. The standard was the principal rallying point for the English army (1).

The *Oxford English Dictionary* (OED) (2), which is our source, cites eight examples of the word used in connection with military or naval ensigns, all of which have documented references. The origin of the sense of the word *standard*, as a "standard of measure or weight" (the ninth OED example) is somewhat obscure. "It is noteworthy that in early instances the standard of measure is always either expressly or by implication called 'the kings standard,' an expression which belongs to the older (military) sense. It seems probable 'standard of measure or weight' is a figurative use of the earlier military standard."

The words *standard* and *standardization* (the action of standardizing, OED) are unique to English. The other European countries use variants of "normal" for "standard"; French *norme*, Spanish *norma*, German *Normung*, Italian *normale*. "Standardization" in these languages becomes some variant of the French *normalization*. Many standards have been agreed upon and accepted by all the countries in the European Union. Such standards are indicated by "EN," for "European Norm," preceding the number designation.

TAXONOMY

In what branch of learning does "standardization" best fit? Science, engineering, technology, or what? To be classified as a science, standardization would be subject to the requirements of the scientific method; that is, a problem would need to be recognized, data collected, a hypothesis formulated, and an experiment designed to test that hypothesis.

The development of such a standard does not follow this scientific course of action, designed to ensure objectivity—a standard is developed when a group of technologists decide to formalize, for example, a procedure or to require a specified level of performance. These decisions are subjectively arrived at based upon the opinions of the technologists and therefore, in our view, make it inappropriate to view standardization as a science.

The dictionaries define *engineering* as "work done by engineers," and since standards are developed by almost every profession (psychologist, biologists, and educators, among many others) it, *engineering*, is not a suitable classification. Technology is also unfitted as a term to describe *standardization* because technology is generally considered as the application of sciences to industrial or practical arts.

What then is standardization? We adopt the suggestion of Verman (3) and classify standardization as a discipline, which Webster defines as "a branch of instruction or learning." Or one can classify it as an appropriate area of study in the less formal sense of a strict academic discipline per se. Either is sufficient for our needs (4).

CATEGORIES OF STANDARDS

A standard, in the simplest sense, is an agreed-upon way of doing something. More elegantly, a standard defines a uniform set of measures, agreements, conditions, or specifications between parties (buyer-seller, manufacturer-user, government-industry, or government-governed, etc.).

Voluntary international standards are one of the major categories of standards and standardization work. "The International Organization for Standardization (ISO) is a worldwide federation of national standards bodies from over 100 countries, one from each country" (5). ISO is not an acronym or abbreviation for the official name. It is a shorthand word derived from the Greek *isos* meaning *equal*, which we find in such words as *isobar, isometric, isotherm*. ISO traces its roots back to 1906 with the founding of the International Electrotechnical Commission (IEC). International standardization continued with the formation of the Federation of the National Standardizing Association with an emphasis on mechanical engineering. All activity stopped at the onset of World War II. Immediately upon cessation of hostilities in 1946 at a meeting involving representatives of 25 countries, the delegates initiated action to create a new international standards organization "the object of which would be to facilitate the international coordination and unification of industrial standards." In February 1947, the ISO came into existence. In collaboration with ISO is the IEC, responsible for developing and harmonizing voluntary international standards in the fields of electrical components, products, systems, and electrotechnology. Some ISO and IEC committees combine their work and thus a reference to ISO/IEC guides or standards.

National standardization is another major category of standards work and standards development. In each of the member countries of the ISO, for example, one organization is chosen to represent it. In the United States it is the American National Standards Institute (ANSI). Whenever ISO undertakes to develop a standard, ANSI designates the appropriate technical advisory group (TAG) of individuals drawn from one or more American standards-writing organizations. For example, if the standard to be developed is related to boilers, the American Society of Mechanical Engineers (ASME) would be a likely choice since they formulate the boiler codes in the United States. The ASME would select from its members one or more experts in the area addressed by the standard under consideration. This person will attend the meetings of the ISO committee or working group as an active, participating member, all the other committee members having been selected by similar procedures to represent their country's (sometimes parochial) interests.

Usually one or more national standards are proposed as possible candidates and the members discuss, argue, modify, and adapt the proposals until tentative agreement is reached. Each member returns to his country and exposes the proposal to his colleagues who may accept, reject, or offer modifications. These are then presented to the ISO committee for further discussion. The process, which may take several years, is repeated until agreement by two-thirds of the participants is achieved, and it becomes an ISO International Standard when 75% of voting members approve.

STANDARDIZATION AND STANDARDS DEFINED

ISO standards are consensus standards but our brief description does not give the reader a full understanding of what that implies. Among many thousands of standards, the ISO has developed definitions for *standard* and *standardization* and related terms that are better suited for our use than those found in the common dictionaries or encyclopedias. ISO's definitions for *standardization* and *standard* are given in the following text.

Standardization

The process of formulating and applying rules for an orderly approach to a specific activity for the benefit and with the cooperation of all concerned and in particular for the promotion of optimum overall economy taking due account of functional conditions and safety requirements.

It is based on the consolidated results of science, technique, and experience. It determines not only the basis for the present, but also for future development and it should keep pace with progress. Some particular applications are:

1. Units of measurements;
2. Terminology and symbolic representation;
3. Products and processes (definition and selection of characteristics of products, testing and measuring methods, specification of characteristics of products for defining their quality, regulation of variety, interchangeability, etc.); and
4. Safety of persons and goods.

Standard

The result of a particular standardization effort, approved by a recognized authority. It may take the form of (1) a document containing a set of conditions to be fulfilled (in French *norme*) or (2) a fundamental unit or physical constant, for example, ampere, meter, absolute zero (Kelvin) (in French *talon*).

The French have distinguished these two forms with the words *norme* and *talon*. For English ISO has assigned *concept* and *term*; *concept* connoting the kind of standard that is the result of an intellectual effort and *term* implying a measured quantity, such as the meter. By far, most standards are written documents frequently called *paper* standards. These are definitions, descriptions of procedures, statements of required results, etc. The lesser number are the physical standards that are at the base of the entire discipline. These were in earlier times the metal bar with two lines engraved on it that defined the meter, the metal cylinder that defined the kilogram, or the vessel that defined the standard for volume. These devices were in the custody of the National Bureau of Standards (NBS) now named as the National Institute of Standards and Technology (NIST). The NBS had been established as a response to the requests of the scientific,

Pascal Krieger. Reprinted with permission from ISO Bulletin, Geneva, Switzerland.

engineering, and industrial community to establish an agency to develop and calibrate standards.

Modern technology requires more precise specification of length than was possible with the standard bar. The new method uses spectroscopic radiation from the rare gas krypton. The meter is now defined as 1,650,763.73 wave lengths of the orange-red radiation of krypton 86 and the U.S. yard as 0.91439980 meter. The older physical standards continue to be used wherever appropriate. A brief history of physical standards is given in "Weights and Measures Standards of the United States," NBS Special Publication 447, U.S. Department of Commerce.

Specification (6)

ISO has also defined another type of standard, *specification*, as follows; A concise statement of a set of requirements to be satisfied by a product, a material, or a process indicating, whenever appropriate, the procedure by means of which it may be determined whether the requirements given are satisfied.

Notes: (1) A specification may be a) a standard, b) a part of a standard, or c) independent of a standard. (2) As far as practicable, it is desirable that the requirements are expressed numerically in terms of appropriate units, together with their limits.

In Note 1, the first two cases describe situations in which the standard describes a test procedure, and stipulates that if the product being tested does not complete the test successfully (as described in the standard) the product is not acceptable. In the third case, the specification calls out a standard and then cites a specific minimum or maximum value the product must achieve to be acceptable.

LABORATORY ACCREDITATION

Industry and government depend on laboratory test results to guide regulatory and corporate decisions daily. It is important that the data be accurate and reliable. One method to assure that the test data has these qualities is through the accreditation of laboratories. Accreditation is defined by ISO/IEC as: "formal recognition that a testing laboratory is competent to carry out specific tests or specific types of tests" (7).

Not all laboratories wish to be accredited. This does not necessarily question their technical competence but shows that they derive no benefit from accreditation. Most U.S. accreditation plans are designed to meet the specific needs of some industrial or governmental program. One set of requirements for laboratory accreditation does not fit all needs. M.A. Breitenberg, National Institute of Standards and Technology, points out, "Some schemes entail only a simple review of data submitted by a laboratory with no attempt at verification. Others require a full scale on-site evaluation of the laboratory's facilities, staff and equipment, and include audits, quality system review, and proficiency testing" (8).

The National Voluntary Laboratory Accreditation Program (NVLAP) is such a program. The candidate testing laboratories run the gamut from industry in-house, federal, state, municipal, and industry association to independent laboratories. Some may engage in a narrow range of testing such as window air conditioners, whereas others may test a variety of electrical appliances. ISO/IEC defines a test as a "technical operation that consists of the determination of one or more characteristics of a given product, process or service according to a specified procedure'' (7).

Frequently, governments require by law that tests to qualify a product be carried out by an accredited laboratory. Foreign governments may insist on the same practice for goods imported into their country. There are other reasons laboratories become accredited:

1. To establish credibility;
2. To show competence;
3. As proof of superiority over nonaccredited laboratories; and
4. As a way of protecting in liability suits.

In recent years, governments are using the term *designation* instead of *accreditation*. *Designation* specifies one or more test houses to carry out tests of products for qualifying those products for agency or public use under some regulation. Designated laboratories are accredited but only those that are designated may be used to carry out tests for some specified purpose.

The validity of the laboratory accreditation process depends on the integrity of the laboratories. They must be independent of all manufacturers, agencies representing manufacturers, or any organization that might have influence to bias their test results. In other words, there shall be no conflict of interest. Breitenberg discusses a number of other criteria for the laboratories (8–9), among them are financial security, on-site inspection, staff qualifications requirements, adequate quality systems (which ensure acceptable performance), sampling requirements, and control and statistical methods.

CERTIFICATION

Certification is that process by which a product or process is assured to meet the requirements of some purchaser. The process may take many forms.

Self certification: The process in which the manufacturer or producer assures the ultimate customer that the product conforms to the label claims. An example is over-the-counter drugs marked with the USP designation, which assures that the product conforms to the specifications of the U.S. Pharmacopeia. Aspirin is a common such substance. Motor oil weights (viscosity) are designated by the Society of Automotive Engineers by the letters SAE followed by a code such as 10W–30W. Alcohol content in hard liquors is fixed by proof (two proof units being equal to 1% alcohol by volume).

Third-party certification: This is ''a form of certification in which the producer's claim of conformity is validated, as part of a third-party certification program by a technically or otherwise competent body other than one controlled by the producer or buyer''.

Numerous organizations and laboratories are third-party testers. Among these are such familiar names as Underwriters Laboratories (UL), Factory Mutual Engineering and Research (FM), American Dental Association (ADA), American Gas Association (AGA), and the Air-Conditioning and Refrigeration Institute (ARI). There are well over a hundred private sector U.S. organizations that engage in certification activities. (For a list see Ref. 8.)

FEDERAL GOVERNMENT CERTIFICATION PROGRAMS

Such government certification programs are essentially of three types:
1. Those that deal with products that affect the public's health and safety, such as those requiring approval of drugs (Food and Drug Administration—FDA); of human and animal drugs and medical devices, etc., (Department of Health and Human Services); of aircraft (Federal Aviation Administration—FAA); and of recreational boat equipment (U.S. Coast Guard—USCG).
2. Those that deal with military equipment. The Department of Defense (DOD) has established a Qualified Products Listing (QPL) of parts and materials used for defense. This program eliminates retesting of qualified products.
3. Those that deal with various food stuffs. Meats, dairy products, vegetable, fruit, and related products by the Department of Agriculture (USDA); and fish and shellfish by Oceanic and Atmospheric Administration (NOAA).

STATE CERTIFICATION PROGRAMS

States are responsible for many programs, some of which are delegated from the federal government. The USDA authorizes states to inspect meat and brand the meat with the USDA stamp of approval. The Department of Housing and Urban Development (HUD) authorizes states to issue certificates approving manufactured homes. The states may impose more stringent requirements than the federal government and states may also prescribe measures to ensure that the produce of their state achieves and maintains a desirable quality. Florida and California do so for citrus fruit, Nebraska does so for tractors, and California does so for auto emissions.

CONFORMATION

In the early 1990s the ISO/IEC subsumed the word *certification* in the word *conformation* or *conformity assessment*. Conformation assures that a product, system,

Pascal Krieger. Reprinted with permission from ISO Bulletin, International Organization for Standardization, Geneva, Switzerland.

process, service, or testing laboratory conforms to a standard. The methods of determining whether the product conforms are the same as for certification, such as

1. By third-party certification systems
2. By manufacturer's declaration of conformity
3. By testing only
4. By inspection only
5. By international systems, etc.

ISO/IEC has issued an extensive manual describing these processes entitled *Certification and Related Activities* (10).

REFERENCES

1. RE and TN Dupuy. The Encyclopedia of Military History. 1971, pp 291.
2. Oxford English Dictionary. Oxford, England: Oxford Press, 1971.
3. LC Verman. Standardization: A New Discipline. Hamden, CT: Archon Books, pp 17–19, 1973.
4. SM Spivak and KA Winsell, eds. A Sourcebook of Standards Information. G. K. Hall Reference. New York: Macmillan. 1991, pp 32.
5. ISO Compatible Technology Worldwide. Geneva: ISO, 1993.

6. TRB Sanders. The Aims and Principles of Standardization. Geneva: ISO, 1972, pp 17–19.

7. International Organization for Standardization and the International Electrotechnical Commission. Guide 2. Geneva: ISO and IEC.

8. MA Breitenberg. Laboratory Accreditation in the United States. National Institute of Standards and Technology, NISTIR 4576, Gaithersburg, MD.

9. MA Breitenberg. The ABC's of Certification in the United States. National Bureau of Standards, NBSIR 88-3821, Gaithersburg, MD.

10. International Organization for Standardization. Certification and Related Activities. Geneva: ISO, 1992.

4

Purposes and Scope of Standardization

In 1689 Boston was destroyed by fire. To speed the reconstruction, the city fathers decreed that the kiln operators produce only 9 in. × 4 in. × 4 in. bricks instead of any of the varieties of sizes they had made in the past. Violation of this mandate resulted in a term in the stocks or pillory. This decree illustrates the purposes of standardization: to establish a routine, to establish a requirement to conform, and to establish specifications and standards (1). With this mandate the city managers insured that the brick makers, the masons, and the architects all knew what to expect with regard to bricks. In addition, it made possible a reasonably accurate prediction of time, manpower, and costs to complete construction of buildings.

This standard and most standards do not hamper innovation or creativity. However, there are situations in which standards can have exactly that effect and precautions should be taken to avoid such results. For example, building codes that specify in detail the construction and materials that may be used in a building may prevent the development or use of new, cheaper, and effective products. Say, for example, that the code specifies thermal insulation to be 4 in. (10 cm) of rock wool sandwiched between sheets of fire-proofed paper 2 mils (0.45 mm) thick. Such a code would prevent any improvement in thermal insulation. One way to avoid this effect is to establish performance standards; that is, to say how much heat may be transmitted across the barrier at a specified temperature differential. Of course, this requires the establishment of a standard test procedure with a specified maximum heat transfer instead of a design specification. It is possible to use performance properties wherever an inherent property such as thermal or acoustic transmission is controlling.

SCOPE

Standardization is a discipline involving many factors. Lal Verman (2) has created a "standardization space" that illustrates geometrically the relations between the several attributes of standards. Verman's space shows the subject on the x-axis, the aspect on the y-axis, and the level on the z-axis. A subsequent elaboration of the level or hierarchy of the standardization space has been developed by Toth (R.B. Toth, personal communication) (Fig. 4.1), which better reflects the complexity of levels in North American standardization. In Toth's expansion of the z-axis, the levels now become the following:

1. International mandatory (or treaty)
2. International voluntary (nontreaty)
3. Regional mandatory or voluntary
4. National mandatory
5. National voluntary

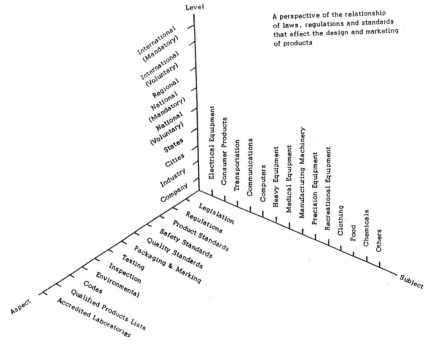

Figure 4.1 Diagrammatic representation of standardization space, a perspective of the relationship of laws, regulations, and standards that affect the design and marketing of products, after L. C. Verman as modified by R. B. Toth.

6. State or provincial
7. Local codes, usually mandatory
8. Professional associations and technical societies
9. Industry or trade associations
10. Company or individual

Subject standards encompass almost all of man's commercial and technological fields; aspect names the forms and types of standards; and level distinguishes the tier in the hierarchy in which the standard operates.

The subjects are each multifaceted, for example, engineering covers aeronautical, civil, electrical, mechanical, hydraulic, fire protection, etc., and science covers biology, chemistry, physics, astronomy, etc.

Verman describes aspect as follows. "Standards differ in form and type depending on the particular aspect of a subject that may be covered. The aspect may be:

"(1) a set of nomenclature or definitions or terms for a given field of industry;

"(2) a specification for the quality, composition, or performance of a material, an instrument, a machine or a structure;

Source: Reproduced from *Through History with Standards*, credit given the American National Standards Institute (formerly American Standards Association).

"(3)   a method of sampling or inspection to determine conformity to a specified requirement of a large batch of material by inspection of a smaller sample;

"(4)   a method of test or analysis to evaluate specified characteristics of a material or chemical;

"(5)   a method of grading and grade definitions for natural products, such as timber minerals, etc;

"(6)   a scheme of simplification or rationalization, that is the limitation of variety of sizes, shapes or grades designed to meet most economically the needs of the consumer. This also includes dimensional freezing of component designs to ensure interchangeability;

"(7)   a set of requirements for packaging and/or labeling;

"(8)   a set of conditions to be satisfied for the supply and delivery of goods or the rendering of a service;

"(9)   a code of practice dealing with design, construction, operation, safety, maintenance of a building, an installation, or a machine, conservation or transport of material or goods, model bylaws, etc., and

"(10) a model for routine use or a model contract or model agreement; and so on."

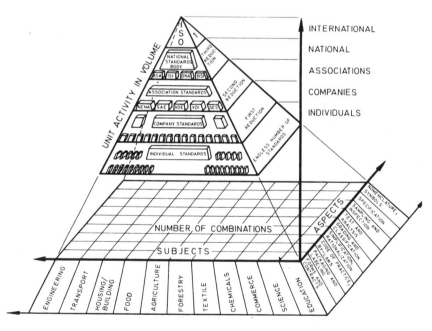

Figure 4.2   Pyramidal representation of standardization space (after Riebensahm), reprinted by Lal C. Verman.

Level specifies the domain in which the standard is operative. Riebensahm (3) has combined the three attributes in a pyramid showing the multiplicity of points in the standardization space (Fig. 4.2). This space has no properties other than presenting in an orderly fashion all the attributes of standards.

THE TIME DIMENSION

Standardization is not a static discipline. Materials are continuously being improved, knowledge and understanding grow and need change. As a consequence, standards must be changed to keep pace with the times. Many standards-writing organizations require that standards be resubmitted to their membership periodically for reapproval with or without changes. If the standard is not reapproved, it is removed and is no longer valid. Typically, reapproval is required every five years. In practice this five-year limit is stringently enforced by certain voluntary standards developers, such as International Organization for Standardization (ISO), American Society for Testing and Materials (ASTM), and National Fire Protection Association (NFPA). By contrast, mandated or regulatory standards (inexplicably) are much slower or more difficult to revise; or the regulation or the issuing agency may not provide the time limit for revision, a "sunset provision," or the personnel for timely revision of their standards.

AIMS OF STANDARDIZATION

The concluding discussion summarizes the excellent 1972 ISO booklet *The Aims and Principles of Standardization* (4). Refer to the original source for more details.

Simplification: Standardization is the means by which society gathers and disseminates information, and disciplines its flow for society's benefit and safety.

Interchangeability: Simplification tends to limit varieties and increase interchangeability.

Standards as a means for communication: An essential function of standardization is to facilitate communication between producer and customer by specifying what is available and giving confidence that the ordered goods will comply with his requirements as stated.

Symbols and codes: By international standardization of symbols and codes, differences in languages are ameliorated or eliminated.

Safety: Safety has two aspects. There are products that have only one function, namely to confer a safe condition on the user. Familiar examples are the safety belt and air bags for motorists; rubber gloves and surgical masks to protect physicians and patients; life preservers; and protective clothing of various kinds. A second aspect of safety deals with uniformity of product. An unduly variable product (item to item or along its length) may fail under conditions of normal use to the peril of the user. Uniformity as used here is a statistical concept that prevents the occurrence of a dangerous condition to some specified limit such as less than one in ten million.

Consumer and community interest: Consumers have a growing interest in durability, reliability, energy consumption, and flammability, etc. These information needs are being met by product labels that report the results of standard tests frequently carried out by certified laboratories (see Chap. 3). Community interest is expressed in various laws, regulations, or codes that protect the entire population. We see these in regulations designed to ensure clean air and water; to prevent electrical fires in homes; and safety performance standards for automobile tires.

Reduction in trade barriers: In recent years the international community has engaged in efforts to eliminate the practice of individual nations from excluding the products of others by imposing unique standards on imported goods that are not essentially different from the product being protected. The General Agreement on Tariffs and Trade (GATT), predecessor to the World Trade Organization (WTO) has dealt with this area of contention and, in large part, eliminated it by making it common practice to accept different standards that produce essentially the same result.

THE PRINCIPLES OF STANDARDIZATION

Principle 1:   Standardization is a conscious act of simplification by the society. It prevents the generation of unneeded variations of products where those variations do not provide any new or unique service.

Principle 2:   Standardization is a social and political as well as an economic activity and should be promoted by a consensus of all concerned.

Principle 3:   A standard is useless unless it can and will be used. Its use may require some sacrifices by the few for the benefit of the many.

Principle 4:   Standards are compromises between various alternatives, involving a decision and to use an agreed upon option for a period.

Principle 5:   Standards should be reviewed at regular intervals and revised or eliminated as desirable.

Principle 6:   When product characteristics are specified, the standard test methods must be designated. When sampling is required, the size and frequency of the samples must be specified.

Principle 7:   The necessity for legal enforcement of national standards should be considered, having regard for the nature of the standard, the level of industrial development, and the laws and conditions prevailing in the society for which the standard has been prepared.

REFERENCES

1. American Standards Association, superseded by American National Standards Institute (ANSI). Through History with Standards. In: R. Glie, ed. Speaking of Standards. Boston: Cahners Books, 1972 (out of print), pp 1–36.
2. LC Verman. Standardization: A New Discipline, Hamden, CT: Archon Books, 1973, pp 33–35. (out of print).
3. H Riebensahm. Report on Initiation and Promotion of Company Standardization among Indian Industries. New Delhi: Indian Standards Institution, 1966, pp. 31.
4. TRB Sanders, ed. The Aims and Principles of Standardization. Geneva, Switzerland: International Organization for Standardization, 1972.
5. ASTM. The Handbook of Standardization: A Guide to Understanding Today's Global Standards Development Systems. West Conshohocken, PA: ASTM, 1999.
6. American Society of Mechanical Engineers. The Why and How of Codes and Standards. . . . New York: ASME, 1984.

5

Genesis of Standards and the Law

Standards are generated by organizations in various ways:

1. By requirement of law—mandatory;
2. By consensus—voluntary;
3. By agreement within an organization—voluntary; and
4. By either number 2 or 3, but with the intent or likelihood that the standard will thereafter be adopted or referenced into codes, laws, or regulations.

The first two have regard for preserving the due process guaranteed by the Constitution's Fifth Amendment although they are safeguarded by different mechanisms. Due process limited to this subject requires that all parties with an interest in the standard will have an opportunity to present their data, their experience, and their opinions to the standard writers; and that these views will be considered and acknowledged. If the objections or recommendations are rejected, the reasons for those rejections will be given.

We will describe the standards-developing methods of the Federal government's mandatory regulations, the American Society for Testing and Materials (ASTM), some safety standards organizations, and the American National Standards Institute (ANSI). These were chosen because they are among the largest producers of standards and ANSI is coordinator of the U.S. federated national standards system.

FEDERAL REGULATION

The first step in the process is the enactment by Congress of an act in which it lays out its objectives, and describes and defines the constraints on rulemaking

33

in some detail. The responsibility for the rule making is delegated to some agency within one of the departments or administrations in the Federal establishment. The personnel assigned to the task begin by studying the legislative history to learn the Congress's intent in passing the act. For example, if the act requires that the standard be stated in performance terms, the intent may have been that the standards be performance standards rather than design standards.

The act usually also requires that the standard tests be reasonable and practicable, which indicates that the responsible official (in many cases the Department Secretary) must consider reasonableness of cost, feasibility, and adequate lead time in the promulgation of the regulations.

The Administrative Procedures Act (1) provides for two types of rule-making procedures—informal and formal. The informal process requires publication of notice of a rule-making action in the Federal Register with the opportunity for interested parties ''to participate in the rulemaking through submission of written data, views, or arguments with or without opportunity for oral presentation. After consideration of the relevant matter presented, the Agency shall incorporate in the rules adopted, a concise general statement of the basis and purpose.'' For the formal rule-making process, a hearing examiner presides at a trial-like proceeding with witnesses undergoing cross-examination. The record of the hearing and evidence submitted in writing becomes the basis for rule making. A detailed description of the informal process is given by Brenner (2).

The informal process is cumbersome. To illustrate, the process begins with an Advanced Notice of Proposed Rule Making requesting suggestions and comments of what should be in the rule. The time at which the comments are due is specified—usually 60 to 90 days. Among the industry comments received are usually at least one suggesting that the agency abandon the effort.

The agency now considers the comments, incorporating those it finds meritorious along with the ideas generated within the agency based on its researches. These provide the basis for a Notice of Proposed Rule Making, which it publishes in the Federal Register inviting comment due at a specified date. Again, the comments are analyzed, accepted, or rejected and incorporated in the revised rule. In the preamble to the rule, the reasons for rejecting any suggestions that were received will be given. The revised rule will be published in the Federal Register as a Notice of Rule Making with the date on which it becomes effective. On the effective date, the regulated industry or any interested party may petition the appropriate U.S. Court of Appeals for review of the regulation with the hope of proving that the agency has not followed the rules or that the standard is not soundly based. If they are successful the court will find that the standard is not valid and reject it.

It should be clear that all interested parties have the opportunity to offer their data and views, to voice their opinions and objections, and to have them considered in the rule-making procedure.

ASTM CONSENSUS STANDARDS-MAKING PROCEDURES

The first step is the perception that a standard is needed. The need may be voiced by any of many sources: ASTM committees, federal or state governments, industry, consumer groups, technical associations, etc. ASTM must accept and act on the request unless another standards-writing organization is already active in the area or if the request is for standardizing an illegal product. This is important because by refusing to take on a task ASTM would be exercising censorship without a consensus being obtained.

ASTM appoints an ad hoc committee to organize a committee to take on the task. An organizing meeting is scheduled and the notice of its time and place is distributed widely to all standards-writing, scientific, and engineering organizations. The participants develop a title, scope, and committee structure to address the tasks it deems necessary to accomplish its goals.

Committee Organization

A committee that will develop standards dealing with materials, products, systems, or services that are offered for sale is required to be *classified*. This means that each individual committee member must be classified according to the following scheme:

Producer: A member who represents an organization that produces or sells materials, products, systems, or services covered in the committee scope.

User: A member who represents an organization that purchases or uses materials, etc., other than for household use provided that the member could not be classified as a producer.

Consumer: A member who primarily purchases or represents those who purchase products and services for household use.

General interest: A member who cannot be classified as a producer, user, or consumer. (Government, academics, and consultants are frequently classified in this category.)

Voting

A company may have any number of its employees be members of a committee but only one of those employees can vote on matters concerning approval of standards. This rule prevents controlling a committee by flooding it with employees from one organization.

Balance

In a classified committee, balance is achieved when the combined number of voting user, consumer, and general interest members equal or exceed the number of voting producer members. This provision assures that the producers cannot dominate the development of standards.

Other Restrictions

1. The chairman of the committee cannot be a producer.
2. When technical matters relating to the development of standards are discussed the meeting must be open to all visitors.
3. Minutes of committee and subcommittee meetings must be taken and sent to ASTM headquarters.

Development of a Standard

1. Standard development begins in a subcommittee with the specification of the need. A task group is assigned the task of developing a draft standard. When the group feels the draft is ready, it is submitted to the Subcommittee for Letter Ballot. A copy of the ballot is filed with ASTM Headquarters. The ballot specifies a return date at least 30 days after mailing.

2. At least 60% of the voting members must return ballots; at least two-thirds of the combined affirmative and negative votes cast by voting members are required.

3. Negative votes *must be accompanied by written statements* explaining the reason(s) for the negative. The reasons given must be technical in nature. Note: only official voters may vote affirmative; but anyone, member or not, may vote negative.

4. Negative votes shall be acted on by the subcommittee in a scheduled meeting or by letter ballot. The negative voters shall be informed as to the time and place of any meeting to consider the negatives.

5. The subcommittee will determine if the negative vote is "not related" to the matter being balloted, or "not persuasive." Approval requires two-thirds of the combined affirmative and negative votes cast by the voting members. If the negative is found persuasive, the draft standard is sent back to the task group for revision or withdrawal.

6. When the draft standard is accepted at the subcommittee level, it is submitted to the main committee for letter ballot by ASTM headquarters. The same procedures govern at this level as applied at the subcommittee level, except that at least 90% of the combined affirmative and negative votes cast by voting members is required. Negative votes are handled by the subcommittee as described above.

7. Upon approval by the main committee, the standard is submitted to the membership of ASTM for approval. Negative ballots are treated as described above.

8. A negative voter who is not satisfied with the decision relating to his negative may appeal to the Committee on Standards (a committee of the ASTM Board of Directors).

To this point we have been dealing with classified committees, those committees and subcommittees that develop standards dealing with materials, products, systems, or services offered for sale. Each main committee has subcommittees that do not fall into this category, such as the executive subcommittee that directs the main committee; the subcommittees on definitions, statistics, and the like; the Society's Board of Directors, its Committee on Standards, its Committee on Technical Committee Operations, and such committees as those on definitions and on statistics are examples of unclassified committees or subcommittees.

This discussion has been extracted from "Regulations Governing ASTM Technical Committees" (3) and does not cover every situation that arises in a subject as broad as this.

LEGAL INVOLVEMENTS

Each and every standard dealing with materials, products, systems, or services restricts in some sense. For example, if a specification "Joint Compound for Finishing Gypsum Board" is approved, widely adopted, and incorporated in building codes in most jurisdictions, one or more producers may find themselves excluded from the market unless they make expensive changes in their production processes. They may feel that their competitors conspired with the standards-making organization to produce the standard to disadvantage them. Over the years such companies have sued, claiming restraint of trade under the Sherman Anti-trust Act. During this century although ASTM has been subject to numerous suits in which they have been cited for restraint of trade they have not once been judged as guilty. They have been protected by their strict adherence to the procedures given above.

In a case that attracted a great deal of attention, the Johns-Manville case (231 F. Supp. 686[1964]) Judge Van Dusen charged the jury:

"Now I rule as a matter of law . . . there isn't sufficient evidence in the case in so far as the American Society for Testing and Materials is concerned . . . to justify your [finding] . . . that the defendants (Johns-Manville) brought about the adoption by the American Society for Testing and Materials specifications designed to increase the costs of foreign-made asbestos-cement pipe and couplings, to render such products ineligible for use, and to otherwise restrict and eliminate competition from such foreign-made products. . . . They might have tried to do it. But they could not, in view of the fact they (the defendants) were only two of 17 members, of some 30 members of this C-17 Committee."

In a separate action (4), Judge Van Dusen entered these findings of fact: "There is no evidence that Committee C-17 or any of its subcommittees were ever dominated or improperly influenced in any way by any of its members. There is no evidence that any member of the Committee ever did more than simply express to his fellow members his scientific viewpoint on the subject of proposed specifications.

"Because of the balance of interests represented on ASTM committees and because of the detailed and scrupulously observed procedure which governs their operation, it is most unlikely that the views of one member or one group of members could predominate over the consensus of opinion of the Committee as a whole.

"The technically qualified, balanced membership of ASTM committees, and the democratic procedure governing their operation, make it likely that the results reached by them will be scientifically sound and will represent the general interest.

"Because the specifications written by ASTM are the result of careful research, open discussion in committees and subcommittees, and thorough review at all levels it is most unlikely that any ASTM specification could ever be passed which constituted an unreasonable restraint of trade."

Thus, it is clear that a standards-writing organization will be free from the accusation of being in restraint of trade if it follows the principles detailed previously: open membership, balance of interests, the adherence to democratic procedures, the dealing with each and every negative ballot, etc.

ANSI AND AMERICAN NATIONAL STANDARDS

What ANSI Is

The ANSI is a private, nonprofit membership organization that (1) coordinates the U.S. voluntary consensus standards system, (2) approves as American National Standards those standards submitted by standards developers that meet ANSI criteria, and (3) serves as the U.S. member body to the International Organization for Standardization (ISO) and the International Electrotechnical Commission (IEC). ANSI was founded in 1918 and is headquartered in New York, NY with a governmental liaison and conformity assessment office in Washington, DC.

ANSI heads the U.S. private-sector standards federation, and is a membership organization comprising about 1,300 member companies, 40 federal government agencies, and about 250 professional and trade organizations representing business interests, technical societies, consumer professionals, trade and labor groups, and others interested in American national and international standardization. As a consensus organization and coordinator of American National Standards, ANSI assures that a single body or set of nonconflicting standards are developed and then approved by ANSI-accredited standards developers. Further, the ANSI approval process sees that all interested and concerned parties have the opportunity to participate in the standards-development process and that due process in standards development has been followed. There are about 9,000 ANSI-approved standards today.

What ANSI Is Not

Due to the diverse and multifaceted nature of the U.S. standards system, both private and public, it is also important to know what ANSI does not do so as to avoid misunderstanding. First, ANSI is not a government organization but is private and mainly funded by the private sector. This is effective but atypical for national standards bodies. For example, of the over 100 members of the ISO, only ANSI/USA and SNZ/Switzerland are fully private with little federal funding. Second, ANSI does not develop standards, per se, rather it establishes the criteria and the approval process for other's standards to be recognized as American National Standards. There is still considerable confusion over this issue of standards development. ANSI used to develop standards in decades past, but has not done so for many years and is precluded from doing so by its current practices. Still, the confusion lingers but the distinction is important and must be emphasized.

ANSI'S OTHER FUNCTIONS

ANSI is a primary source for information on and access to both American National Standards and those published by the ISO, IEC, and other standards bodies with whom it holds sales agreements. Regional standards developers such as those in the European Economic Area (EEA, formerly EU or EC) provide standards proposals and final drafts for comment. These are announced by ANSI and available for purchase.

The ISO Compendium on international quality management standards is available for purchase by ANSI as well as from ISO and other sources, as are its individual parts such as the ISO 9000 and ISO 10000 series quality standards. Of recent and keen interest are the new ISO 14000 series environmental management standards (EMS), and these standards can be obtained from ANSI.

One manner in which ANSI informs its membership and the public of national and international standards developments is through its monthly *ANSI Reporter*. This provides a timely management report of issues germane to strategic standardization, activities of its standards developers, relations with the government sectors, and more. ANSI functions through a series of councils, committees, and boards, among which are the Company Member Council, Organization Member Council, Government Member Council, Consumer Interest Council, U.S. National Committee to the IEC, and more.

ANSI Approves American National Standards

There are three means by which ANSI reviews and approves American National Standards that are submitted by some standards developer. These are:

1. The accredited organization method—in which the overall organization's standards-development procedures, due process, and consensus comply with ANSI's requirements;
2. The accredited standards committee—in which a standards committee (but not necessarily its parent organization) is following the required ANSI procedures; and
3. The canvass method—which involves a privately generated standard that is submitted to ANSI, who canvasses all interested and materially affected parties to be notified in writing or informed (thus canvassed) of the impending process and standard under canvass review. If all criteria for process and approval are met, the standard is given an ANSI designation and published as an American National Standard.

The reader will have noticed that the ANSI standards must be developed insuring that the due process and consensus procedures that are required have been met. We present here a case in which these requirements were not met, with disastrous results to the sponsoring organization. The case is known as *Hydrolevel v. ASME* (456 USC 556 [1982]) (5). Briefly, the Hydrolevel Company had developed a device to be attached to boilers that sensed when the water level fell below the safe level. In order to market this device the approval of the American Society of Mechanical Engineers (ASME) was required. ASME is responsible for administering the boiler codes and its ruling are law; no boiler equipment can be used in the United States that does not have their approval. Hydrolevel was unable to obtain that approval. It turns out that one of Hydrolevel's competitor's volunteers to the approving committee conspired with another volunteer to prevent approval of the device. Hydrolevel was forced into bankruptcy but continued to fight in the courts and eventually won a multimillion-dollar settlement. The case was appealed to the Supreme Court by ASME claiming that they were not responsible for the action of volunteers. However, the Court found the injury was inflicted by a false interpretation of the boiler code administered by ASME. The Court stated:

> "It is true that imposing liability on ASME's agents themselves will have some deterrent effect, because they will know that if they violate the antitrust laws through their participation in ASME, they risk the conse-

quences of personal civil liability. But if, in addition, ASME is civilly liable for the antitrust violations of its agents acting with apparent authority, it is much more likely that similar antitrust violations will not occur in the future.''

In summary, Hydrolevel was denied "due process" by not being present when critical decisions were made and further "consensus" was not achieved since the decisions were made in secret by a cabal of two. It was one of very few landmark cases for the standards system.

OTHER STANDARDS DEVELOPERS IN THE UNITED STATES AND CANADA

There are at least 250 standards developers in the United States of which ASTM is only one, and about 400 or more if you count the minor organizations and those with only a few published standards. The majority of published U.S. standards and standards developers are coordinated by ANSI within the American national standards system. See Chapter 14 for a complete analysis.

This is a highly diverse, healthy, and competitive standards milieu. It is somewhat odd or confusing to others in the world but not to those knowledgeable or familiar with its operation. This structure is comprised of numerous large and small standards developers, of which ASTM is one. There are many other than ASTM, some noted in the following text, and their standards services are vital to those within their sphere of influence.

NFPA, UL, AND OTHER SAFETY STANDARDS DEVELOPERS

The National Fire Protection Association (NFPA), an organization promoting fire safety in North America and worldwide, is best known for its development of the Life Safety Code and for a large body of fire-protection and life-safety standards. Underwriters Laboratories (UL) is one of the premier standards developers for electrical and safety features of components, products, and systems. It should be noted that UL does not "certify" products in the traditional sense, but rather "lists" components and products that are tested and meet its safety standards. The UL listing or equivalent mark is often incorporated into legislation as a requirement on products or systems sold and used in local or state jurisdictions, building codes, and the like.

The National Electrical Manufacturers Association (NEMA) and the American Gas Association (AGA) are industry trade associations that develop harmonized product performance, labeling, and safety standards for their respective industry's products. The American Water Works Association (AWWA) and the National Sanitation Foundation (NSF) do the same for water quality, safety, and usage.

STANDARDS ENGINEERING SOCIETY (SES)

The United States and Canada have a professional association of individual standards personnel interested in the standards development process and in sharing that knowledge among their membership. The SES (Miami, FL) serves the needs of standards neophytes to standards experts and technical-standards information professionals from business, industry, government, and academia. SES is a source of professional credentials through its standards engineering certification program; publishes the bimonthly *Standards Engineering* magazine; and holds an annual conference on topics of concern vital to standardizers, standards developers, information specialists, and many others.

One of the key roles of SES is to provide a network among standards professionals through regional chapters in the United States and Canada, and internationally as well. SES also serves as the North American member to the International Federation for the Application of Standards (IFAN).

SES Standard on Standards

A recent contribution by SES was its development of a standard on standards. In 1989, SES initiated the development of a standard to address the needs of standards users in attempting to identify and locate existing standards and to provide uniform guidance on the format and designation of standards for standards developers. An SES committee was established to be responsible for the development of this standard and included representatives of major standards-developing organizations (e.g., ARI, AWWA, ASTM, NFPA, SAE, NISO; see Appendix A), federal government agencies (e.g., DoD, DoE, GSA, NIST), major corporations (e.g., Eastman Kodak, General Electric, Boeing, Deere & Co., Fluor Daniel), and other interested parties.

The ANSI/SES standard entitled "Recommended Practice for Standards Designation and Organization" (6) establishes recommendations for the designation of standards, standards titles, the use of abstracts and keywords in standards, and a uniform format for the contents of standards. It also provides recommendations for identifying revised portions of standards in both draft revisions and approved revisions of standards.

The SES standard identified as "SES 1" has also been approved by ANSI as "ANSI/SES 1: 1995." Copies of the standard may be obtained from SES headquarters, 13340 SW 96th Avenue, Miami, FL 33176, United States or from commercial standards sellers.

DUE PROCESS AND REASONABLENESS IN STANDARDIZATION

In summary, it is worth recalling what the basic elements or principles of due process are that apply in ASTM, to ANSI for approval of American National Standards and in many other standards organizations. The following is excerpted from the ASTM Committee on Standards ByLaws.

"1. Timely and adequate notice of a proposed standards undertaking to all persons likely to be materially affected by it.

2. Opportunity of all affected persons to participate in the deliberations, discussions and decisions concerned both with procedural and substantive matters.

3. Maintenance of adequate records of discussions and decisions, with timely publication and distribution of minutes of meetings of all committee meetings.

4. Adequate notice of proposed actions.

5. Meticulously maintained records of drafts of proposed standards, action on amendments, and final promulgation of standards.

6. Timely and full reports on results of balloting.

7. Careful attention to minority opinions throughout the process, and adequate means for procedural review and appeals."

Further, the U.S. Federal Trade Commission has reviewed the reasonableness of standards and proposed some factors to be considered in judging same (see "US Federal Trade Comm. Staff Report on Standards and Certification"). They are as follows:

"1. The purpose of the standard must be legitimate, reasonable and clearly shown, socially desirable and in the public interest.

2. The requirements of a standard for a product or process shall be those which can be reasonably met by all segments of the industry and should be generally acceptable to users and consumers.

3. The standard shall be written, if possible, in the broadest performance concepts to encourage innovation and invention and to promote technology.

4. The standard shall not be written in such a way that it can be used to mislead users or consumers of the product, service or process covered by the applicable standards.

5. Test methods required by the standards should be reasonable and adequate to measure the characteristics in question. The needed personnel and equipment to conduct the tests should be generally available and at reasonable cost.

6. Provisions involving business relations between buyer and seller should not generally be included in a standard, except as it pertains to determining conformity assessment.

7. Certification, accreditation, conformity assessment and quality assurance markings or requirements must be reasonable and not restrictive.

8. No standard shall be written which requires the use of a patent, trademark or service mark unless such patents or ownership are available on a non-discriminatory basis, free of charge for a reasonable fee.''

REFERENCES

1. Administrative Procedures Act, 5 USC, Secn. 553.
2. FC Brenner. The Uniform Tire Quality Grading Standard. ASTM Standardization News 10 (March), 1982.
3. ''Regulations Governing ASTM Technical Committees,'' West Conshohocken, PA: ASTM (revised periodically).
4. MR Brooke. Protecting Standardization Activities From Antitrust Problems: A 1986 View. In: Materials Research and Standards, 1968, pp 10–19.
5. HM Markman. Hydrolevel vs. ASME. ASTM Standardization News 11 (July), 1983. Reprint NFPA Journal 77, no. 2 (March), 1983.
6. U.S. Federal Trade Commission. Rulemaking Hearings and Reports on Standards and Certification: Final Staff Report, April, 1983. See also report of the presiding hearings officer, June, 1983.

6

Varieties and Relationships Among Standards

In this chapter we discuss the varieties among standards, their relationships, and applications. We also report on the efforts of the federal government to reduce the needless duplication in standards development and usage, by adoption of federal policy for the development and use of voluntary (nongovernmental) standards. In particular, we summarize the Office of Management and Budget's OMB Circular A-119 (1998 revision in Appendix B) and simultaneously, by incorporating this standards policy as legislation in Public Law 104-113, the National Technology Transfer Improvements Act of 1996 (NTTIA or TTIA).

DESCRIPTION OF STANDARDS

Standards are designated by descriptors that usually tell how the standard was formed, who developed it, its purpose, and date of last revision. There are a variety of standards with differing uses and applications. Often the types of standards may be directly contrasting, in conflict, or competitive; consider the following.

Governmental standards are in most cases mandatory or regulatory standards and developed in what is known as the public sector. They are most times opposite in form and function than are nongovernmental or voluntary standards. Mandatory standards are just what their name implies and include both federal purchase and procurement standards of the U.S. General Services Administration (GSA) implementing the Federal Acquisition Regulations (FAR) or those of the U.S. Department of Defense (DoD). If you are in business and wish to provide goods or services to GSA or DoD then you must comply with all of their applicable standards and specifications.

Mandatory purchase specifications also apply at the state, provincial, or local level, for example, states such as California, New York, Texas, Illinois, and many others have large numbers of product specifications. Chapter 7 gives an example of how procurement standards may be wisely developed.

Regulations exist with the force of law and will often contain or reference applicable standards. Mandatory or regulatory standards are found throughout federal, state, and local laws and regulations, for example, health and safety standards, environmental standards, trade regulation rules, labeling requirements, and others. Codes and standards are either written by, adopted, or referenced into legislation and become mandatory or prescriptive standards. At the local level, city or county, state or province, applicable codes such as the Life Safety Code, Uniform Building Code, and plumbing and electrical codes are mandatory standards within their jurisdictions. This is described in more detail in the following text. We have shown in Chapter 5 how a governmental or regulatory standard is created and the safeguards in force that preserve the rights of the regulated industry while others, such as users and consumers, are protected.

By contrast, voluntary standards are not required to be used or complied with unless the buyer and seller, or interested and affected parties, choose to so use them. Many voluntary standards are written by the consensus method and following due process procedures such as in the American Society for Testing and Materials (ASTM). Or they may become American National Standards or ANSI standards if they meet the ANSI 'tests' of due process in standards development. Voluntary standards are the domain of some 250 major standards developers in the United States and Canada; and up to 400 or more counting smaller standards developers. This decentralized system (described earlier in Chapter 5) is in sharp contrast with most other countries of the world where one standards developer predominates. Or even Canada where there is one standards coordinator, the Standards Council of Canada (SCC), and only a handful of accredited standards developers (e.g., CSA, CGSB, ULC, BNQ).

Industry trade associations typically develop and publish standards for their specific industry, such as the National Electrical Manufacturers Association (NEMA), Association of Home Appliance Manufacturers (AHAM), American Petroleum Institute (API), and others. Customarily, these standards are developed by industry experts and not necessarily in a consensus mode.

Professional and technical societies are those most familiar to standardizers and the largest producers of voluntary standards. In Chapter 5 we have shown how a voluntary consensus standard is developed and how integrity of the process is assured by conducting open meetings, by maintaining unrestricted membership, by maintaining ''balance'' in the committees, and guaranteeing that every disagreement (negative vote) is resolved with due process assured to the dissidents. These standards developers operate in a more open and exposed environment and many utilize consensus and due process to produce their standards. One of the best recognized is ASTM, one of the world's largest consensus standards developers. Others include the Society of Automotive Engineers (SAE), the American Society of Mechanical Engineers (ASME code and standards), the National Fire Protection Association (NFPA codes and standards), American Society of Heating, Refrigerating, and Air Conditioning Engineers (ASHRAE), and hundreds of others.

Standards developed by professional societies, technical or trade groups, or trade associations may ultimately become ANSI standards bearing the designation ANSI/XXXX-year. They may be developed by an accredited standards organization, accredited standards committee, or be submitted to a written canvass review process. If acceptable, and in compliance with ANSI procedures and review, they will be considered as future American National Standards.

Certain independent or private-sector standards developers also must be mentioned. These standards developers initially develop their own standards, but many achieve ANSI or SCC status in time. Premier among these is Underwriters Laboratories (UL) whose standards and product listings for public safety are used worldwide. At the international and regional levels, an additional set of descriptors or terms apply for the varieties and applications of standards. There are literally hundreds of regional and international standards developers and it is important to note that these too are distinctive as producing either mandatory or voluntary standards. If mandatory then there will be some treaty of legally binding covenant between signatory nations incorporating the standard. At the international level, mandatory standards are best seen in the work of the new World Trade Organization (WTO) headquartered in Geneva, Switzerland and its predecessor the General Agreement on Tariffs and Trade (GATT). One of the many companion signatory treaties to the WTO and GATT is the Agreement on Technical Barriers to Trade, popularly known as the Standards Code. The WTO has also instituted the General Agreement on Trade in Services (GATS) and the recently approved Information Technology Agreement (ITA), which have received attention because of the burgeoning issues of standardization of services and cross-border trade in information technology.

Another mandatory standards treaty is that of the International Organization of Legal Metrology (OIML) headquartered in Paris, France. OIML maintains the standards of reference and agreements comprising the international system of units (SI), sometimes referred to as the modernized metric system of measurement and commerce. The United States is among only two or three countries of the world that have not legally adopted the SI or metric system as our primary unit of weights and measures.

At the regional level, mandatory standards are most seen in European Norms (EN) or standards of the European Union (EU). In particular, it is those standards that are incorporated into EU Directives that become mandatory and thus carry the force of law and strict compliance. Examples of regional mandatory standards are those referenced in evolving EU Directives such as the Low Voltage Directive and other directives for medical devices, worker health and safety, machinery, toy safety, and the like. Other regional standards are those of the North Atlantic Treaty Organization (NATO), whose equipment interoperability standards assure that the respective parts are interchangeable. Other regional standards organizations include the Pan American standards organization known as COPANT and headquartered in Caracas, Venezuela; the Pacific Area Standards Congress (PASC) that includes not only the Asian Pacific nations but also the United States and Canada; the Gulf Cooperation Council (GCC) whose secretariat is the Saudi Arabian Standards Organization (SASO), and several others.

Voluntary international standards are best represented by the three companion organizations in Geneva, Switzerland whose international standards predominate in world trade and commerce. These include the International Organization for Standardization (ISO) approaching 10,000 published standards; the International Electrotechnical Commission (IEC) with 2,500–3,000 standards and related documents; and the International Telecommunications Union (ITU). Others of note are the Food and Agriculture Organization of the World Health Organization (FAO/WHO) in Rome, Italy whose Codex Alimentarius (or Codex) of international food additives and food safety standards are used worldwide.

CODES

Codes are a special class of standards of particular interest to state and local authorities having jurisdiction (AHJs). *Webster's Third New International Dictionary* defines *code* as "c: a set of rules of procedure and standards of material designed to secure uniformity and protect the public interest in such matters as building construction and public health, established usually by a public agency and commonly having the force of law in a particular jurisdiction [a building code] [changes in the sanitary code]."

Professional and industry associations have taken responsibility for developing codes. The American Society of Mechanical Engineers (ASME) develops the Boiler and Pressure Vessel Code, and the elevator codes. The National Electrical Code and the Life Safety Code are developed by and the responsibility of the NFPA. Among other industry organizations involved in code development are the National Sanitation Foundation (NSF), and the three model building-code organizations namely the International Conference of Building Officials (ICBO), the Southern Building Code Congress International (SBCCI), and Building Officials and Code Administrators (BOCA); their umbrella organization, the Council of American Building Officials (CABO), and now the International Codes Congress (ICC), who represents the national code consensus and promotes the model codes. There are several other code related organizations (1).

Depending upon who is doing the counting, it's variously estimated that there are on the order of 4,000 primary code jurisdictions in the United States and somewhere between 10,000 to 40,000 total jurisdictions with code authority and their own AHJs. Given the several competing code authorities noted in the preceding text and the lack of a uniform national code in the United States, it became apparent that some harmonization of this disparate problem was in order. The International Codes Congress (ICC) was established in 1993 with the aim of developing a uniform set of national codes by the year 2000. A model plumbing code has already been completed; a national electrical code for one- and two-family dwellings is near completion; and ongoing plans culminating in a national fire code by the year 2000 are in the works.

GOVERNMENT AND THE VOLUNTARY STANDARD SYSTEM

For many years, some of the agencies of the federal government generated the standards they needed. Two agencies, DoD and GSA, published volumes of standards covering their purchasing needs. Other agencies contributed their share of government standards in areas of occupational safety and health, environmental protection, consumer safety, transportation, and many more.

About 1970 an interest developed to see what could be done to make certain the federal standards writers took advantage of the expert knowledge in the voluntary consensus system and to assure that the government's standards were not duplicating efforts in the private sector. In 1968, an effort originated within the Department of Commerce (DoC) by its forming the Interagency Committee on Standards Policy (ICSP), which continues to serve as a coordinating body for federal government standards policy; some 30 federal agencies and departments take part in periodic meetings of the ICSP. The committee formulated a number of elements around which a policy should be formed (2). Among these were the directions that federal agencies should:

1. Participate in outside standards-setting activities when they are in the public interest;
2. Reach agreement with the nonfederal or nongovernmental standards bodies that meetings will be open and nonrestricted; and
3. Use voluntary standards in lieu of in-house standards when they serve the public interest and are economically as or more desirable.

The work of the ICSP after much interagency debate and comment by ASTM, ASME, and ANSI was formalized. In 1976, a draft of the revised policy was sent to the OMB with a request that it issue a formal OMB Circular prescribing federal policy on the matter. After further review and comment, OMB Circular A-119 on "Federal Participation in the Development and Use of Voluntary Standards" was first issued in 1982. Ten years later in 1993 and again in 1998, the Circular was revised and republished (3). This Circular's purpose is to establish "policy to be followed by executive branch agencies in working with voluntary standards bodies. It also establishes policy to be followed by executive branch agencies in adopting and using voluntary standards."

In brief summary, A-119 charges federal standards developers and standardizers to first explore what existing voluntary standards may be available and applicable for their use in the government's need. If such standards do not exist or are not currently applicable then the federal agencies are instructed to coordinate and cooperate with private-sector standards developers to see that the necessary standards can be developed in a timely fashion to meet the government's requirements. Federal standardizers are also encouraged to participate in and support the development of private-sector standards where it is in the public sector's best interest. In the absence of any of these alternatives working or solving the government's standards need, then the agency and its standardizers may develop their own government standards as they see fit.

Circular A-119 did not have the force of law, it being effectively policy guidance for executive branch agencies only. It was followed by some agencies but ignored by others. K. Kono observed (4), "In a 1995–1996 study conducted by the National Research Council (NRC) of the Academy of Science and Engineering, it was reported that 'Current efforts by the US government to leverage the strengths of the private US standards development system as outlined in OMB Circular A-119 are inadequate. Effective, long term public-private cooperation in developing and using standards requires a clear division of responsibilities and effective information transfer between government and industry. Improved institutional mechanisms are needed to effect lasting change.' " A notable exception in recent years has been the continued reduction in the number of DoD defense purchase and procurement standards in favor of more simple commercial item descriptions (CIDs) as well as adoption of nongovernmental standards where applicable.

"On March 7, 1996, President Clinton signed into law . . . 'The Technology Transfer Improvements Act of 1995' [TTIA] and contained in the new law is a provision (12(d)) that codifies the existing OMB Circular A-119 on 'Federal Participation in the Development and Use of Voluntary Standards.' " Time will tell whether this new law will be recognized or even followed any more than OMB Circular A-119; see Appendix B for this OMB Circular.

REFERENCES

1. AE Cote, CC Grant. Building and Fire Codes and Standards. In: AE Cote, ed. Fire Protection Handbook. 18th ed., Quincy, MA: National Fire Protection Association, 1997, pp 1-42–1-54.
2. FE Clarke. The Genesis of OMB A-119. In: W. Conshohocken, PA: ASTM, and the OMB Circular A-119. May 18, 1983, pp 13–16.
3. OMB A-119. Federal Register 63, no. 33 (19 February 1998):8546–8558. OMB A-119, Federal Register 58, no. 205 (26 October 1993):57643–57648.
4. KR Kono. OMB A-119 Becomes Law. ASTM Standardization News. May 1996, pp 40–42.

7

Procurement Standards

This section is reprinted with permission and minor editing from the Standards Council of Canada Symposium brochure by Norm Hagan titled "Implementing a Standards Program for Procurement." Norm Hagan is President of Norm Hagan & Associates, Inc., management consultants in Westport, Ontario, Canada.

INTRODUCTION

Although there are many publications available extolling the virtues of using standards in procurement, very little is in print explaining how to put standards into practice. This publication is intended to help fill that void. If you have the responsibility or the opportunity to set up a standards program, this booklet is for you. It will also be useful to purchasing officers and other managers, to increase their understanding of their roles in such a program.

Several important considerations should be kept in mind when using this information. The ideas are intended to serve only as guidelines. You must tailor your program to the requirements of your organization. A standards program is a support service to others, and not an end in itself. For the vast majority of the items you purchase, standards already exist. Try to find such information. Do not reinvent the wheel.

A brief word on terms and definitions. There is some confusion regarding the use of the terms *standard* and *standardization*. For the sake of clarity, in this chapter a *standard* means "a written document defining the requirements for products or services, and adopted for use on a recurring basis." The term *standardization* includes the establishment of specific requirements, and the approval of products or services to meet those requirements. It also includes strategies necessary to effect savings by the selection of standard products and services, promotion of their use, and simplification of their procurement. Standardization permits economies of scale and minimization of an unnecessary variety of products that perform the same function.

WHERE SHOULD YOU START?

If your organization has decided to embark on a standardization program, it is probably already convinced of the advantages. Some of these may be:

1. The economies to be realized;
2. The need to obtain improved and consistent quality of products purchased;
3. Ensuring fair and competitive bidding through the use of generic specifications rather than product trade names; and
4. Greater probability of long-term availability of products.

Whatever your organization's objectives may be, the question arises as to where to start. What standards are now in use? The area you should research is your own organization. If you are not certain whether standards are already used by your purchasing group, speak to the purchasing officers. It may be that some of the departments for whom they purchase already provide standards or specifications. If so, obtain copies and become familiar with them. Some of those documents may be useful to other departments.

Management Commitment

Management commitment and involvement are vitally important, especially in the initial stages. Commitment to the standards program, as well as involvement in the setting of policies, will ensure that management has the opportunity to use its knowledge and experience to influence the direction the program will take. At the same time, managers will become aware of their roles and the responsibilities their staff will assume. One way to do this is to form a standards steering committee, consisting of the top management people from each section or department, to make decisions on policy matters. You should draft proposed policies and procedures and submit them to this committee for discussion and approval.

PLANNING

As with almost any project, good planning is the key to success, and that means having the right people involved in the process. The right people will be those with sufficient authority and responsibility to implement the program. This involvement is necessary to ensure that your plan fits the management style of your organization, and to have the commitment of the people who will make it work. If you can get the plan right the first time around, you will gain credibility much more quickly. At this stage of the planning process, you should place more emphasis on the people aspect of the program and less on technical matters.

The plan you develop should include all of the activities to be carried out. It should show who is responsible for the completion of each activity, as well as a time schedule. Be sure that the plan is agreed to and responsibilities are accepted for the tasks involved. Once you have a workable plan, you will have set the scenario for implementing the program.

Policies and Procedures

One of the key areas in your planning will be the development of policy and procedures. Does your organization have a written policy for procurement stating that the award of contracts for goods and services will be made on the basis of fair competition? Does it mention that evaluations of tenders will be made on the basis of conformance to the minimum specified requirements for the goods and services to be supplied? If not, you should develop and have approved a statement that explains clearly and briefly your organization's policy with regard to standards. Some typical statements might read as follows:

1. It is the policy of (organization) that one of the criteria used in the evaluation for the award of contracts will be the supplier's ability to fulfill the requirements of any standards or specifications that form a part of such contracts.
2. It is the policy of (organization) that standards or specifications will be used for procurement purposes, wherever appropriate, in order to ensure impartial and competitive bidding.

A further concern will be the processes used for procurement in your organization. If you are not thoroughly familiar with the methods, documents, and procedures in use, get to know them. A related item to consider is the policies and procedures governing procurement in your organization. They may exist in written form or be adopted practices. It is particularly important that new policies and procedures you propose are not in conflict with those already in existence, but support them. Management is in the best position to provide judgement on such matters. At this time, it would be wise to prepare a summary paper on your findings. This can be used for the next step, which is planning your program. This document should summarize the results of your research and should recommend a course of action for your organization.

Once you have an approved policy statement, you can develop procedures to implement that policy in a practical manner. The procedures to be considered should include:

1. Who is responsible for standards development;
2. Who is responsible for the approval of standards;
3. What will be the general methodology for the development or adoption of standards;
4. Who should be consulted or participate in the development of standards;
5. Who is responsible for administering the various activities of the standards program; and
6. Who will be responsible for quality assurance and any testing activities.

These and other realms of responsibility should be considered so that the role of everyone who may be involved in the standards program is clearly defined and understood. Once approved, these procedures should be incorporated in your organization's policy and procedures manual.

THE STANDARDS PROGRAM CONCEPT

Before discussing standards documents, it will be useful to look at the overall picture of a standards program. What you should visualize is a closed loop. This concept is analogous to the cycle of activities that may take place in the standardization process. The principle elements are shown with their relationships in Figure 7.1 and described in the following text.

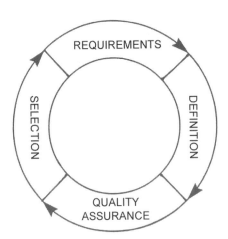

Figure 7.1 Diagrammatic representation of the procurement standards development process.

1. Determination of the user's requirements (research)
2. Definitions of the user's requirements (adopting or writing a standard)
3. Verification that the purchased product meets the requirements (quality assurance)
4. Tendering and subsequent product selection (acquisition)

Remember that having a standard is a means to an end. That end is consistently to provide the user of a product with a quality suitable for the work to be performed, at the most economical end cost. Whenever you are considering adoption or publication of a standard, you must also be concerned with how it will be applied. This application could be in the determination of compliance for a tender award, or verifying that you are indeed receiving an acceptable quality product.

GENERAL METHODOLOGY

Having determined that a recurring problem in the purchase of a particular product might be resolved through the use of a standard, you should ask several preliminary questions. Is the problem, as defined, actually caused by the lack of a standard? Will having a standard eliminate, or substantially reduce, that problem? If the answers are *yes*, then some or all of the following steps should be considered:

1. Clearly identify the problem.
2. State performance or design requirements necessary to resolve the problem.
3. Search for an existing relevant standard or
   a. Modify an existing standard to meet your requirements.
   b. Draft a new standard, with input from all persons directly involved in the problem.
4. Obtain information from suppliers to ensure availability of the product, as specified.
5. Have the standard approved by purchasing personnel and the end users.
6. Implement the standard by referencing it in purchasing contracts.
7. Monitor implementation of the standard to ensure the desired results are obtained. Involving people in the preparation of a standard gives support to the resulting standard.

ADOPTING STANDARDS

Having decided that a standard is needed and having determined what the requirements for the product are, you should endeavor to find an existing standard that covers those requirements. You should begin your research using the indexes available from one of the recognized standards-writing organizations. If you are unable to find a suitable standard, another possible avenue to explore is the Standards Information Service of the Standards Council of Canada (SCC). They maintain a computerized database from which they can make a search for existing standards by topic, keyword, or other criteria. The Standards Information Service will also be able to tell you who to contact for information on government jurisdictions or the private sector.

If you find a standard that seems to suit your requirements, or nearly so, you may wish to take an additional step in order to confirm its applicability. Obtain a sample of a product manufactured to the standard. It will be much more meaningful to users if you show them a product made to a standard than for them to try to interpret the technicalities of that standard. This method can be used for those types of products where the physical characteristics are obvious. It is somewhat more difficult in the case of products such as chemicals or paints, where performance may be a criterion to be judged. Nevertheless, if it can be shown that a well-known and proven product is made to the standard, then in most cases you can have confidence in your selection of that standard.

In some instances, you may find that a standard covers some provisions not pertinent to your requirements. In such cases, it may be possible to adopt only those portions of the standard that meet your requirements. For example, only the "Test Methods" of the standard may be applicable. In any case, you should consider adopting only those standards that are developed and maintained by a recognized standards-writing organization. Doing so will permit you to monitor any changes made to that standard, thus ensuring that it continues to fulfill your requirements.

It is important to ensure that the standard being adopted or written is acceptable to the end user. If the equipment is being exported, purchasing components to North American standards may not be acceptable. In this regard, it should be noted that ISO (International Organization for Standardization) and IEC (International Electrotechnical Commission) standards are becoming widely accepted both overseas and in the United States and Canada.

Legislated Standards

You must at all times be aware of standards that have become law. Most legislated standards are related to safety, health, and welfare. The various acts and regulations in your jurisdiction will inform you of the existence of such laws. In addition, it is well to be aware that certain federal legislation governs products manufactured, imported, or offered for sale within states or provinces. It is therefore vital to ensure conformance to such legislation when purchasing these products.

Approaching the Purchasing Officer and Product End User

When you introduce your program to the people with whom you will deal the most frequently, don't expect to be received with open arms by everyone. Some will look at standards as just so much additional work. Others may not have had exposure to using standards and are not aware of the benefits. In any case, it is crucial that you are not seen to be forcing them to accept your ideas. You will have to "market" the concept of using standards in procurement.

Marketing can best be carried out by helping someone to solve a problem. Seminars or discussions with individuals will very often result in the identification of problems that may be resolved through the implementation of a standard. Initially, the problem may not be thought of in the context of the lack of a standard. Recurring poor quality, improper selection of a product for its intended use, damage owing to poor packaging, and an excessive number of purchase orders for essentially the same item are some typical cases that can be turned into selling points for the idea of using standards and standardization.

Some people will be sold much more readily than others. Work with those persons at the outset. Ask for their advice. Involve them in deciding where standards would be useful, as well as in the development of those standards. They can help tremendously, if you invite their participation. You must be, and be seen to be, a service to purchasing officers and their clients.

WRITING YOUR OWN STANDARDS

Although you may have to write your own standards, try to do so only as a last resort. Every standard you publish is costly in terms of preparation and maintenance. When it is necessary to write your own, you should publish them in a professional and consistent format. The following list of typical contents of a standard is provided as a guideline. It is not recommended that you adopt this format per se. Investigate the formats used by other organizations, then decide on one that suits your application.

Heading: The heading of your document should state the name of your organization, the title of the document, its number, and date of issue.

Scope: A very brief statement of what the standard covers.

References: If you make reference within your document to any other standards, they should be listed, showing the name of the organization that publishes the standard and its reference nomenclature.

Product qualification: Outline any requirements for the qualification or prequalification of products.

Information to be supplied with bidders or tenders: Indicate whether suppliers are required to furnish any information, samples, or documentation at the time of bid or tender.

Definitions: Define any unusual terms, or terms that are subject to wide interpretation.

General: Detail any technical information related to product requirements.

Construction: When applicable, specify details concerning construction.

Tests and test procedures: Indicate any tests or methods that should be used in order to determine conformance to the standard.

Product identification: Indicate product identification, manufacturers' nomenclature, or other pertinent information requirements.

Packaging and package marking: The proper packaging and marking of containers is an important requirement and should be outlined.

Quality assurance provisions: In this section, you can indicate any inspection or other quality-control options that may be exercised or required. Also, indicate any action that may be taken in the event of nonconformance to your standard.

Drawings: Any drawings that form a part of the standard should be included, or referred to, in this section.

Helpful Hints

1. An important point with any standard is to have the revision status clearly identified. Most organizations use a revision letter after the identifying number.

2. When changes are made, indicate the lines or paragraphs affected by a vertical line in the left-hand margin.

3. Be specific and give proper dimensions and tolerances. The use of terms such as *typical* or *nominal* give no control.

4. Keep wording simple and precise. Show information in a tabular form where possible.

5. Ensure that metric measurements are in accordance with the appropriate metric standards.

6. When expressing the date in numeric form, use the ISO sequence YYYY-MM-DD (e.g., 1997-12-25) in accordance with international standards. If most computer programmers had done so in the latter decades of the twentieth century we wouldn't have had a Y2K problem.

SUPPLIER CONSULTATION

There are several reasons for fostering good relations with suppliers and in particular, manufacturers. They can provide you with a great deal of useful information on their own products as well as the industry that produces them. Make use of these resources in determining your requirements. Additional reasons for encouraging good supplier relations are fairness and impartiality. If you implement a standard without prior consultation, which inadvertently eliminates one of your suppliers from qualifying on a tender for a product, you will invite criticism of your organization.

One way in which you can ensure that suppliers have the opportunity to provide valuable technical information and other comments is to circulate selected standards to them before they are actually in use. You then can take their comments into consideration when finalizing your standard. Whenever possible and appropriate, deal with manufacturers rather than distributors. If you follow this approach, you will find that the majority of suppliers will welcome the opportunity it presents to them. In addition, you will help eliminate negative reactions to the introduction of standards.

QUALITY ASSURANCE
What Is Quality?

There is only one meaningful definition of quality. Quality is conformance: conformance to requirements, conformance to a standard. You cannot measure quality in any other terms. A product or service either meets your requirements, fails to meet them, or exceeds them. Once that principle is accepted, you have established the base (your standard) for determining whether a product meets your quality requirements.

What Should You Do About Assuring Conformance?

You cannot control quality; that can only be done by a manufacturer or contractor. You can, however, do something about obtaining quality. The most positive action you can take is to insist that requirements are met when they are specified in a contract. There are many methods of achieving conformance. All of them have application under differing circumstances. It is a matter of selecting the correct option to fit a particular situation. Because it is neither practical nor desirable to inspect most of the items you purchase, other methods should be used in the majority of cases. In any case, a fault detected in a product after manufacture occurs too late to be corrected. Inspection should only be considered as an operational audit.

Some of the various options that can be considered are discussed in the following text. Keep in mind that none of these options can be used unless you have specified your requirements through the use of a standard. That standard can be a document, a comprehensive description, or it may simply be a sample.

Qualified Product Listings

Products that have qualified as conforming to published standards are listed by some organizations. Generally, these listings are issued by standards-writing organizations. However, listings are available from other groups as well. Products appearing on these lists will have met established criteria—either through tests by an independent organization, or through self-certification by a manufacturer.

Certification Statements

You can demand certification from your suppliers or contractors to ensure that their products or services will consistently meet your standards. This is probably one of the best methods to consider, since it puts the onus on the supplier to provide the products you want. This may include the provision of certified or other data related to their offerings. The supplier, then, is more likely to read your standards.

Certification of Manufacturers and Contractors

This method is generally used where large expenditures are to be incurred. In such cases, a recognized certification agency will audit the quality control and reporting methods of a manufacturer or contractor to determine whether they meet prescribed standards. Upon qualification, the manufacturer or contractor will be licensed to advertise and "mark" his product or service under that license. Some of the best known marks in Canada are Canadian Standards Association (CSA), ULC, Canadian General Standards Board (CGSB), and Canadian Gas Association (CGA), and in the United States are UL, American Gas Association (AGA), and Association of Home Appliance Manufacturers (AHAM). Costs of these programs are borne by the manufacturers or contractors.

Formal Inspection and Testing

Inspection and testing certainly have a place in the overall scheme of things; at times it will be the only option. However, large-scale continuing-inspection programs are not recommended in the majority of cases. It is too expensive, not practical, and in any case should be regarded as the supplier's responsibility. Nonetheless, testing is useful when used as an audit, or in special circumstances. Audits may be considered as spot checks and special circumstances may be large contracts, new products, or a new supplier. Deficiencies found on inspection should be handled thoughtfully. The deficiency may be an oversight or an unwitting error. Consultation with the supplier may lead to resolution without rejection or penalty. Do not, however, knowingly accept a deficient product and neglect notifying the supplier.

There are a variety of methods of inspection. They can be carried out by your own organization, by contract to a company specializing in inspection, or to a limited extent, by the end user of a product.

Samples

Retention of samples can be useful in certain circumstances, such as when a standard has not been used for the purchase of a product. In other cases, you may wish to retain samples of chemical-type products. Samples obtained from the successful bidder during the tender award process may be retained, in case of subsequent problems. Should such problems arise, the original sample may be used for comparison analysis against current receipts.

EXPANDING YOUR SERVICES

Productivity Equals Service

Once your policies and procedures are in place, and understood, and you have successfully completed several standards projects, you should consider expanding your sphere of work. Talk to your purchasing officers, user departments, and particularly your warehouse management. Try to ferret out problems they are having that could be rectified by a standard. You should be able to handle more than one project at a time because, under normal circumstances, there is waiting time during the development of standards. This waiting time may consist of arranging for people to meet, awaiting the arrival of information or documents, or for other legitimate reasons. This waiting time can be put to good use on other standardization work.

Advertise Results

Undertake projects where success can be measured in terms of projected dollar savings, increased user satisfaction related to quality, or increased productivity on the part of the purchasing staff in carrying out their work. Use the results of these successes to reinforce your program.

Other Related Activities

As you progress, you will find there are many other ways in which you can be of service. Some of the possibilities are (1) standardized descriptions for catalogues or computerized item description files; (2) standard contracts for services; and (3) promotion of the use of standards in other departments.

AUTOMATION AIDS AND REFERENCE MATERIALS

Word Processing

If you are going to be writing some of your own standards, you will find a word-processing facility indispensable. During the drafting of these documents, or in keeping them up to date, the convenience of a word processor will outweigh the cost many times over. This is due to the number of changes that occur from first draft to final document.

Another way to save time when preparing documents is to maintain on your word processor file most of the standard phraseology used in your documents. This will drastically reduce the time required to write draft copies. You simply provide your word processing operator with the paragraph numbers you wish to use, and they will be called up from the word processor file and inserted automatically in your document. In word processing jargon, this is known as *boiler plating*.

Making Use of Computers

If your purchasing system is computerized, and if you catalogue item descriptions, you should make reference to your own standards or those you have adopted in those descriptions. By doing so, you can be assured that each time a tender is issued, the item descriptions will automatically include those references.

Coding of Standards and Files

You will need to index the standards that you develop or reference. If you use a commodity code system, you should consider using that code as an integral part of your standards numbering system. Consider setting up your standards filing system using the same code. By doing so, you will be practicing standardization and simplifying your work at the same time.

Reference Material

You will require a library of reference material, including standards and specifications from other organizations. Depending upon the scope of your work, this may be costly—but justifiable. You may wish to consider making this material

Adapted by Pascal Krieger for ISO Bulletin, reprinted with permission from ISO, International Organization for Standardization, Geneva, Switzerland.

available to other departments, such as engineering groups. An existing library facility may be the place to keep this material, provided it is conveniently located.

In conclusion, you now realize there are many details to be considered and acted upon in setting up a standards program. Research, planning, and execution involve consultation with many people. Although a standards program might be thought of as being primarily technical, it is the ability to work with people that brings success and personal satisfaction.

REFERENCES

1. TRB Sanders, ed. The Aims and Principles of Standardization. Geneva: International Organization for Standardization, 1972.
2. International Organization for Standardization. Benefits of Standardization. Geneva: International Organization for Standardization, 1982.
3. PB Crosby. Quality Is Free. New York: New American Library, 1979.
4. RP Preston. Standardization Is Good Business. Ottawa: Standards Council of Canada, 1977.
5. AL Batik. The Engineering Standard, A Most Useful Tool. Ashland, OH: BookMasters/El Rancho.

8

National and International Standardization

NATIONAL STANDARDS ORGANIZATIONS

Although standardization is not an invention of the twentieth century, its institutionalization at the national level started in most industrialized countries at the beginning of the last century. The British Standards Institution (BSI) was created in 1901, the predecessor of the Deutsche Insitut für Normung (DIN) in 1917, the American National Standards Institute (ANSI) in 1918, and the Association Francaise de Normalization (AFNOR) in 1926. Like many other organizational structures, the standardization process reflects the specific institutional patterns in different countries, in particular with regard to its integration into the country's general legal and economic framework and its preference for centralized or decentralized structures.

National standardization systems of International Organization for Standardization (ISO) member countries correspond more or less to the following basic models ranging from an extremely centralized approach (vertical model) to a system characterized by the coexistence of either several or many different standardization bodies and the absence of a central focal point for coordination (horizontal model). Intermediate models are systems with a varying degree of centralized authority. See the report on *Consumers, Product Safety Standards and International Trade* (1), as the original source, from the Organization for Economic Cooperation and Development (OECD).

The Vertical or Monolithic Model

Some nations have a monolithic government authority and a centralized governmental standards system. The standards bodies are an integral part of the government, and the national standards organization determines all standards policy. This policy and the application of national standards are communicated down to member branches or regional-municipal bureaus of standardization. The chain of command, control, and standards development is clear and unequivocal. Examples of such vertically oriented standards systems include those in most of the eastern European countries including the former Soviet countries, China, and many of the developing countries.

The Centralized Model

Another system of standardization is one that is characteristic of many developed nations. In these cases, there are strongly centralized national standards bodies and structure, which may be government agencies, or a number of private or quasigovernmental bodies closely linked by structural, contractual, or financial arrangements to the central government. Of OECD member countries the German standards system with DIN at its top or Japan with Japan Standards Association (JSA) is perhaps closest to a centralized structure.

The Decentralized Model

The decentralized model represents more diversity in standards development. There is a clear separation between the voluntary sector and the government sector. The Canadian system is a transitional structure midway between the European centralized model and a broader, more horizontally structured system (2). In Canada, the Standards Council Canada (SCC), a Crown corporation with direct federal funding, accredits the standards-writing activities of the Canadian Gas Association (CGA), Canadian General Standards Board (CGSB), Canadian Standards Association (CSA), Underwriters Laboratories of Canada (ULC), and Bureau de Normalization du Quebec (BNQ). Simultaneously, there are standardization activities by both the Canadian federal government and the equally strong provincial governments. Other testing and certification organizations are also accredited (2).

It may be argued that the United States equally adheres to a national standards approach represented by such a model, although it is even broader at its base than other national standards systems. Coordinating private standards in the United States is ANSI, which approves standards as American National Standards and acts as the official representative to ISO and the International Electrotechnical

Commission (IEC). However, there are perhaps 400 standards bodies writing voluntary standards in the United States, and only about one-half of them participate within the ANSI system. There is a government-standards presence as well, coordinated by the National Institute for Standards and Technology (NIST) and its Secretariat of the Interagency Committee on Standards Policy (ICSP). But akin to the voluntary sector, there are numerous government-standards activities that include 50 states and thousands of municipal code and regulatory bodies. A recent detailed description of standards setting in the United States can be found in *Global Standards* (3).

The Horizontal Model

In this case, the coordination either is not as comprehensive as with the other standards models or has strong coequal partners. The system's base of standards developers and users is extremely large and diverse. This applies to both the private (voluntary standards) and the government-standards sector. Conventional wisdom would place the U.S. national approach into a decentralized model. One may also argue, however, that the U.S. national approach is perhaps better appreciated by analogy to a horizontal system. One striking feature of the U.S. system is the coexistence of a number of private standards developers who are as well known and internationally respected as ANSI, which theoretically has the central coordinating function. Hundreds of smaller standards developers, third-party certification, and laboratory accreditation programs lie outside of the national standards system.

TREATY AND NONTREATY INTERNATIONAL STANDARDS

Treaties are regional or international agreements whereby the signatory nations agree to abide by all aspects of said treaty. There are so-called *treaty organizations* and standards that derive from organizations whose signatories or national member bodies are bound to comply with and adhere to any of its standards. Examples are the World Trade Organization (WTO), successor to the General Agreement on Tariffs and Trades (GATT), and their subsequent treaty agreements, the WTO's General Agreement on Trade in Services (GATS) and the Agreement on Technical Barriers to Trade (known as the Standards Code). Another example is the International Organization for Legal Metrology (OIML) managing the International System of Units (SI) or metric system of units, and many others.

By contrast, a nontreaty agreement is voluntary in nature, so that member organizations or others adhere only on a voluntary basis. Such standards are developed and promulgated voluntarily, although within certain countries their standards may be referenced or adopted into national laws and regulations, a fairly common practice. Examples include the international voluntary standards developers such as ISO and IEC, which will be discussed in the following text.

As an overview to regional and international standardization, the reader needs to recognize that there are literally hundreds of international organizations involved with various aspects of standards and standardization (4,5). Some of these are treaty or mandatory standards developers and others are voluntary or nontreaty standards developers such as the ISO and IEC.

ISO AND INTERNATIONAL STANDARDS FOR A GLOBAL ECONOMY

The ISO (6), founded in 1946 in Geneva, Switzerland (still its headquarters), celebrated its fiftieth anniversary in 1996. ISO is the nontreaty international body that develops, coordinates, and promulgates voluntary standards to help foster world trade, promote quality, protect the health and safety of consumers and the public, protect the environment, and provide aid in technical standards information on a global basis. ISO was first introduced in Chapter 3, in the section on terminology, wherein ISO has an important role in defining a plethora of terms related to standards and standardization.

It should be mentioned that more often than not the reader will see ISO incorrectly named as the "International Standards Organization," a common error as it matches its abbreviation but alas is not its proper name. There is a full description in Chapter 3 under standards terminology to elucidate the relation between ISO, the prefix iso-, and the correct name, International Organization for Standardization.

With the emergence of global strategic standardization as a driving force for multinational business and competition, the rubric or name and roles of ISO (and of IEC) to business and industry have become increasingly important and recognized by CEOs, business, and government leaders (3,7–12).

ISO has published over 10,000 international standards and its growth rate has been increasing steadily. It is perhaps best known for its series of ISO 9000 quality-management system standards, and now its ISO 14000 series of environmental management standards (EMS). Such standards are but a small fraction of the total compass of ISO's overall work and global impact. ISO's main effort is standards writing by almost 200 technical committees and over 2,000 subcommittees in most fields of endeavor. Customarily a national member holds the secretariat and thus has the management role for each technical committee. ISO standards are published and available as final approved documents, or in some cases as draft international standards.

ISO Structure

ISO is made up of national member bodies, of which ANSI is the official U.S. member. In Canada the SCC serves that role. Other well-known national members are, for example, BSI in the U.K., DIN in Germany, AFNOR in France, Saudi Arabian Standards Organization (SASO) in Saudi Arabia, JSA in Japan, and so on. There are 105 or more member bodies of the ISO and they also comprise the ISO General Assembly. In addition to the 200 or so technical committees, there are four policy-development committees of the General Assembly that serve to foster international standards policy. These are:

1. CASCO: the Conformity Assessment Committee;
2. COPOLCO: the Consumer Policy Committee;
3. INFCO: the Information Committee; and
4. DEVCO: the Development Committee for developing nations issues.

The main technical work of ISO (and for that matter of IEC) is done through its technical committee (TC) and subcommittee structure. Key among these is TC 176 on quality management standards (QMS), which has written the ISO 9000

Adapted by Pascal Krieger for ISO Bulletin, reprinted with permission from ISO, International Organization for Standardization, Geneva, Switzerland.

and 10000 series of quality-management standards, and TC 207 on EMS in the ISO 14000 series.

ISO 9000, ISO 14000, and Other Management System Standards

Although ISO has published about 10,000 standards on topics from asbestos to foods to zinc, it is most widely associated with the body of international management system standards. These include the ISO 9000 series of standards for QMS, now in their third generation in 2000–2001. See Chapter 12 for an experiential discussion of compliance with ISO 9000 and 14000.

In the United States, the equivalent national standards are ANSI/ASQC Q9000 series; certain ISO 10,000 series standards, terms, and definitions; and ISO Vision 2000, etc.; these and all other relevant documents are collected in the ISO Compendium of International Standards for Quality Management (13), available from ANSI, American Society for Quality (ASQ), or ISO. For additional information see the references for selected books on ISO 9000 (14–19).

The ISO 14000 series of standards for EMS are also published with ISO 14001 being the first available in completed form. Other ISO 14000 series EMS standards are following in rapid succession. This is another potentiality for ISO

An example of what to avoid: this is unacceptable because it gives the false impression that ISO 9000 is a product quality label. ISO 9000 certificates are issued for quality management systems, not products.

Adapted by Pascal Krieger for ISO Bulletin, May 1997, reprinted with permission from ISO, International Organization for Standardization, Geneva, Switzerland.

consideration for developing international, mutual recognition agreements for quality systems and possibly EMS accreditation programs. An Internet World Wide Web site and resource for national and international ISO 14000 news, training and implementation information, company database, and more has been set up by ANSI and other partners. It can be found at http://www.iso14000.org.III online. It is available on a subscription basis. Many handbooks and references are now appearing on ISO 14000.

Currently under review within ISO is a possible work program that would hope to develop international management system standards for occupational health and safety (OH&S), based on similar precepts to those that underlie the ISO 9000 and 14000 series standards. Although a controversial issue, it's possible that this work will go forward under ISO or other auspices.

IEC AND INTERNATIONAL ELECTROTECHNICAL STANDARDIZATION

IEC plays a counterpart role in international standardization to the ISO, but the IEC's function is to develop and promulgate voluntary international standards in the area of electrical, electrotechnical, and some electronics standardization. IEC is headquartered at the same Geneva, Switzerland address as is ISO, and there are increasing cooperation and coordination between these two bodies of international standardization, and more recently the International Telecommunications Union (ITU) as well. The United States and Canada are represented at the IEC through their respective national committees, for example, where the U.S. National Committee to the IEC is administered by ANSI through which its delegates serve. The IEC has some 2,500 to 3,000 published standards and addenda, and functions through about 90 technical committees and their subcommittees in a manner generally similar to the ISO. Some international guideline documents are jointly published as ISO/IEC guides.

INTERNATIONAL TELECOMMUNICATIONS STANDARDIZATION

The former consultative committees coordinating international standards for television, telegraph, radio, and the like, under rubrics Consultative Committee on International Telephone and Telegraph (CCITT) and Consultative Committee on Radio (CCIR), have been integrated within the ITU. Recently, the ITU in Geneva has joined with ISO and IEC in a high-level tripartite coordination of joint president's policy and in international standardization.

WORLD TRADE ORGANIZATION AND TECHNICAL TRADE BARRIERS

The WTO is successor to the GATT. The GATT has several signatory or binding agreements, the most germane of which, for our discussion, is the Agreement on Technical Barriers to Trade, also known as The Standards Code. First instituted in 1979–1980 with the GATT Tokyo Round, and subsequently updated through the Uruguay Round, The Standards Code stipulates provisions to insure that standards and technical documents should not be used as barriers to international trade.

Of particular importance in the standards code are equal treatment or non-discriminatory (nonpreferential) treatment to all applicable standards affecting goods and services in commerce. Further, The Standards Code states that when nations are developing technical standards and regulations, they should utilize to the greatest extent possible any and all relevant international standards that may exist (ISO, IEC, ITU). Nations can, however, utilize unique national standards in areas such as health and safety, if such can be justified. A complaints and resolution mechanism is established in the case of trade conflict arising over technical standards.

In 1995 with the conclusion of the GATT Uruguay Round, the WTO was established as the formal successor to the GATT and poses more stringent requirements upon its member nation signatories. Included among these are adherence to the WTO's Agreement on Technical Barriers to Trade and included therein is the Code of Good Practice for the preparation, adoption, and application of standards.

The WTO, as well as its predecessor in the GATT standards code, requires that there be national inquiry and information points, which obligates members to disseminate full and timely information on standards development, technical regulations or laws, conformity assessment including certification procedures and programs among all standards code member bodies.

A newer development is the WTO's GATS. This program seeks to develop international agreements so that national standards or regulations shall not be used to hinder burgeoning global trade in services. Since cross-border trade in services is growing at several times the rate of trade in goods or products, this increased attention on standards for services is a very important new development for experts in both national and international standardization.

REFERENCES

1. Organization for Economic Cooperation and Development. Report of the OECD/ CCP Committee on Consumer Policy. Consumers, Product Safety Standards and

International Trade. Paris, France, 1991. Available from OECD bookstore, Washington, DC.

2. SM Spivak, KA Winsell, eds. A Sourcebook of Standards Information: Education, Access and Development, G. K. Hall Reference. New York: Macmillan Publishing, 1992.

3. U.S. Congress. Office of Technology Assessment. Global Standards: Building Blocks for the Future, TCT-512. Washington, DC: U.S. Government Printing Office, March 1992.

4. M Breitenberg. Directory of International and Regional Organizations Conducting Standards-Related Activities, NIST SP 767. U.S. Department of Commerce, 1989.

5. American National Standards Institute. Global Standardization Reports. New York. (Available periodically.)

6. International Organization for Standardization. ISO Bulletin, ISO 9000 News, and numerous articles on ISO and international standardization in respective issues. Geneva, Switzerland: ISO.

7. American National Standards Institute and U.S. World Standards Day Committee. Series of special sections in: Fortune. International Standardization: the Cornerstone of Global Competitiveness. November 2, 1992. Global Connections: The Quest for Worldwide Standardization. November 4, 1991. In: Business Week. American Competitiveness: Gaining an Edge through Strategic Standardization. October 6, 1995. Special issue insert of October 21, 1996.

8. American Society for Testing and Materials. ASTM Standardization News. Several theme issues on international standardization, standards and trade, and conformity assessment. See Conformity Assessment issue. August 1996. Standards and Competitiveness. December 1995. Global Standardization. June 1991. West Conshocken PA: ASTM.

9. International Standards. Financial Times. Survey, 13 October, 1995.

10. National Research Council. Standards, Conformity Assessment and Trade into the 21st Century. Washington, DC: U.S. National Research Council, 1995.

11. JS Wilson. Standards and APEC: An Action Agenda. IIE Policy Analyses in International Economics 42. Washington DC: Institute for International Economics. October 1995.

12. RB Norment. Conformity Assessment and International Trade. Report for the U.S. Consumer Product Safety Commission. Falls Church, VA: Norment and Associates, 1993.

13. International Organization for Standardization. Compendium of ISO 9000 International Standards for Quality Management. Geneva, Switzerland: ISO. (Updated periodically.)

14. J Lamprecht. ISO 9000: Preparing for Registration. New York: Marcel Dekker, 1992.

15. J Lamprecht. Implementing the ISO 9000 Series. New York: Marcel Dekker, 1993.

16. RW Peach, ed. The ISO 9000 Handbook. Fairfax, VA: CEEM Information Services, 1993.

17. F Voehl, P Jackson, D Ashton. ISO 9000: An Implementation Guide for Small to Mid-Sized Businesses. Delray Beach, FL: St. Lucie Press, 1994.

18. RM Smith. The QS-9000 Answer Book. Red Bluff, CA: Paton Press, 1996.

19. H. deVries. Standardization: A Business Approach to the Role of National Standards. New York: Kluwer Academic/Plenum, 1999.

9

Conformity Assessment: Laboratory Accreditation, Quality System Registration, and Product Certification

This chapter was contributed by Frank Kitzantides, Vice President, Engineering with the National Electrical Manufacturers Association (NEMA) of Arlington, VA, United States. It was originally presented at the ISO/CASCO Workshop on Conformity Assessment, Geneva, Switzerland in May 1993 and updated for this book in 1998. His remarks are indicative of electrical manufacturers in the United States, representing the collective wishes of this industry sector.

INTRODUCTION

The last several years, there has been an increasing marketplace and regulatory emphasis on mechanisms for assessment of conformity to requirements specified in standards. The roles of product testing, certification, quality system registration, and laboratory accreditation have become increasingly more important to the U.S. national competitive position as industry more frequently encounters such requirements in its domestic and foreign markets.

In addition, international and regional agreements are bringing new pressures to bear on the manner in which the United States conducts its own assessments of conformity utilizing a variety of approaches specific to a particular industry sector. The increasing role of conformity assessment as a national competitiveness issue; the growing number of sectoral and local or U.S. national certification and accreditation programs; and the potential impact of international standards and agreements on the traditional U.S. approaches to conformity assess-

ment has led the American National Standards Institute's (ANSI) constituents to call for the institute to expand its programs to address these issues.
The principal goals of ANSI with regard to conformity assessment are:
1. To provide a policy forum to represent U.S. interests at the domestic, regional, and international levels (addressing such issues as the mutually support-ive roles of the public and private sectors; cooperation and coordination with U.S. government representation of U.S. interests in multilateral and bilateral trade negotiations; U.S. access mechanisms to foreign conformity assessment systems, and role of suppliers declaration vis-à-vis independent third-party assessment ac-tivities);
2. To provide and assure full use of the U.S. access mechanisms to the private-sector international and regional organizations that set requirements on conformity assessment programs and operating such programs (e.g., International Electrotechnical Commission (IEC) and International Organizations for Stan-dardization (ISO));
3. To collect and disseminate on behalf of its constituents timely, accurate information on conformity assessment developments nationally, regionally, and internationally; and
4. To provide a private sector–based national accreditation mechanism for conformity to provide assessment programs (product certification, quality sys-tem registration, and laboratory accreditation) that facilitate sectoral approaches to satisfy U.S. needs for products and services to flow freely in the marketplace (domestic, regional, and international).

ANSI'S CONFORMITY ASSESSMENT PROGRAM

Accrediting Certification Programs

The ANSI program for accrediting certification programs was established in the 1970s to meet a need for national recognition of competent certification programs, as a private-sector mechanism for self-regulation and market-value enhancement, and for a mechanism to be used by government agencies for purchasing and regulatory-compliance purposes. While historically the majority of U.S. products have been backed by a manufacturer's declaration of its products' conformity to particular requirements, use of third-party certification has been found particu-larly appropriate in the health and safety areas. This trend is continuing as public and environmental health issues intensify. The ANSI accreditation program in-cludes an initial assessment and continuing surveillance of a certification program in order to assure compliance with ANSI criteria.

More and more frequently, U.S. suppliers have encountered foreign certifi-cation requirements as a condition of market access. Domestic suppliers' interest in using United States–based certification programs to demonstrate conformity with foreign requirements is growing. This has led to the desirability of increasing the worldwide acceptance of product certifications performed in the United States and to the promotion of reciprocal agreements between U.S. and foreign certification organizations. The use of national accreditation systems to assess the competence of such certification programs is also growing as a factor in establishing the equivalency of certification programs around the world.

The ANSI policies and procedures for accrediting certification programs have been recently revised to be consistent with relevant international standards and guides and expected obligations for conformity assessment programs under the Uruguay Round of General Agreement on Tariffs and Trades (GATT) negotiations and the North American Free Trade Agreement (NAFTA). ANSI is also working cooperatively with the accreditation systems in Canada and other nations to establish the commonality of the accreditation processes with the goal of completing private sector-to-private sector mutual-recognition agreements.

In addition, ANSI has offered its accreditation program as a tool for U.S. government use in establishing competence of certification programs when this may be required as an aspect of government-to-government negotiations of mutual recognition as in the case of products subject to European Community (EC) Directives that mandate compliance with certain health and safety requirements. The ANSI Accreditation Committee is responsible for the operational aspects of the accreditation program for certification programs. The committee assists in the development of programs relating to certification, reviews the process of program evaluation, and recommends accreditation and monitoring programs of certification activities.

ISO 9000 and ISO 14000 Programs

In response to the growing significance of the ISO 9000 series of quality standards in both the private marketplace and regulated industrial sectors, ANSI and the Registrar Accreditation Board (RAB) of the American Society for Quality Control established the American National Accreditation Program for Registrars of Quality Systems. This joint activity of ANSI and RAB provides U.S. industry with a recognized program for accrediting registrars of quality systems. The program uses internationally accepted guides and procedures as the basis for national accreditation of the registrars so that the industrial clients of accredited registrars are assured of their competence and that the quality system registrations will be accepted in the U.S. and global marketplace.

In recognition that international acceptance of U.S. registrations to ISO 9000 series standards will be a key to continued U.S. competitiveness, high priority has been placed within the program on cooperation and dialogue with counterpart accreditation activities in other countries and internationally. This effort is intended, ultimately, to lead to mutual-recognition agreements between the American National Accreditation Program and other nationally based accreditation systems. The program is also positioned to serve as the ''competence demonstration'' element relating to quality system assessments when this may be required as part of any formal government-to-government negotiation of mutual-recognition agreements for regulated products. Work is now underway to extend the program to incorporate the ISO 14000 standards for environmental management systems (EMS).

International Electrical and Electronics Certification, IECEE/IECQ

Through the U.S. National Committee to the IEC, a committee of ANSI, the United States is actively participating in the IEC System for Conformity Testing for Safety of Electrical Equipment (IECEE), and the IEC Quality Assessment System for Electronic Components (IECQ). Both programs are striving to reduce the number of duplicate tests and certifications around the world by promoting the acceptance of valid test results among the participating member body countries.

International Conformity Assessment Committee, ISO/CASCO

Through ANSI, the United States is actively participating in ISO/CASCO. The ANSI International Advisory Committee is responsible for the U.S. participation in the ISO Council Committee on Conformity Assessment (CASCO). To achieve the aforementioned goals, ANSI has established and undertaken the following additional activities. Aspects of the increased ANSI attention to conformity assessment include the following issues.

Establishment of BCCA

The establishment of a Board Committee on Conformity Assessment (BCCA) in 1992 provides the single primary focal point for coordinating, developing, and maintaining ANSI's appropriate activities addressing conformity assessment issues. The committee makes policy recommendations related to conformity assessment to the ANSI board and provides oversight for ANSI's operational programs.

In 1992, the ANSI board directed the BCCA to initiate an American National Accreditation Program for Laboratories and authorized it to conduct those appropriate activities to define a national program that would be responsive to the sectoral needs identified. The BCCA has established the objectives for an ANSI laboratory-accreditation coordination program, with recognition that there may now exist one or more operational laboratory accreditation programs whose roles should be considered and evaluated in relation to ANSI activities. The Task Group on Laboratory Accreditation will prepare a plan for ANSI's role in laboratory accreditation for consideration by BCCA.

Meetings with EU and EOTC

Supporting frequent meetings and dialogue with the private sector European Organization for Testing and Certification (EOTC) and the European Union (EU) regarding U.S. access to the European market. Conformity assessment to European requirements will be a significant factor in many instances for individual product sectors subject to EU's directives. ANSI's dialogue with the European Commission and the EOTC has produced timely and relevant information for U.S. businesses.

Educational Events

The sponsoring of national educational events to raise general awareness of conformity assessment issues and domestic implications of international trends is encouraged.

Accreditation and Quality System Registration

Staff and resource support for the accreditation program for certification programs and a national program for registrars of quality systems and EMS in partnership with the RAB. The three elements of conformity assessment from the point of view of the U.S. electrical manufacturing industry are expanded upon in the following text.

1. The product certification system, which provides assurance that a product conforms to an accepted standard;
2. The quality system registration, which provides a mechanism that assures the competence of a supplier's quality system; and
3. The laboratory accreditation system, which promotes confidence in the accuracy of the data on which decisions are made.

THE U.S. PRODUCT CERTIFICATION SYSTEM

The U.S. approach to product testing and certification is a pluralistic one. Many private-sector organizations (including trade associations, independent laboratories, professional societies, building code organizations, and standards development organizations) are involved in or have some responsibility for product testing and certification. The growing importance is a natural outcome of purchasers' increased concern for product safety; the need for additional assurance of compliance in the face of potential product-liability litigation; and expanding regulatory requirements at both the national and local levels.

Certification services (by a third party) are generally available on a voluntary basis for a variety of product types. However, the percentage of manufacturers seeking such certification can vary significantly by product area. For some electrical products, those normally used in the home for example, gaining product approval from an independent testing and certification body is generally an industry practice. The situation is different for other electrical products, such as a large electrical apparatus or one-of-a-kind products. Many of these items are not required to be tested or listed by an independent third party. In the case of electrical materials and products used in construction, third-party testing and certification is usually required by state or municipal building authorities.

The degree of confidence that can be placed in any third-party certification program varies greatly depending on: (1) the adequacy and appropriateness of the standards used in the program; (2) the number and types of testing and inspection methods used within the program to assure product conformance; (3) the adequacy of the manufacturer's quality control system; and (4) the competence of the body that conducts the testing or inspection and evaluates the test results.

At the federal level, there are a number of different requirements for testing and certification programs that affect electrical products. The Mine Health and Safety Administration (MHSA) for example, certifies electrical equipment used in mines. The Food and Drug Administration (FDA) has requirements for radiation-emitting electronic products, such as television receivers, x-ray equipment, and ultraviolet lights, and also regulates electrical and electronic devices used in medical applications. The Federal Trade Commission (FTC) has energy-efficiency standards and labelling requirements for many major appliances. The Occupational Safety and Health Administration (OSHA) regulates the safety of electrical products used in the workplace and has established an accreditation program of evaluating laboratories' competence in providing certification programs for products under the OSHA jurisdiction. The Consumer Product Safety Commission (CPSC) collects information about consumer product complaints and hazards and can take action to protect consumers from unsafe products.

At the state and local levels, the installation of electrical products is enforced by state and municipal authorities based on the requirements of the National Electrical Code (NEC), which is adopted as a mandatory code. Over 40 states have adopted the NEC without deviations. Others have minor-to-moderate modifications based on local conditions. States and municipal authorities also generally require the approval of electrical products and materials prior to their placement in installations subject to inspection. Most jurisdictional authorities accept product testing and listing by nationally recognized testing laboratories as being indicative of the product's acceptability. A few states have even established criteria for the accreditation of such laboratories (e.g., North Carolina, Texas, Washington, Oregon).

Obviously, the decision to implement certification may be motivated by regulatory requirements, customer expectations, liability considerations, or a combination of these factors. The choice between a manufacturer's declaration and certification may be influenced by these same factors. Also influencing a decision on whether to use a manufacturer's declaration is the existence of (or need for) competent test and technical capability. The manufacturer will have to consider such factors as availability and facilities, trained and experienced personnel, and suitability of in-house testing standards.

NEMA and ANSI and their constituency believe that a manufacturer's declaration is an option that should continue to be made available to manufacturers who will determine its viability and applicability to their individual situations. Those responsible for product approval (such as inspection authorities) should carefully weigh and consider this option.

QUALITY SYSTEM REGISTRATION

There is also a continuing trend by U.S. manufacturers to register quality systems involved in the production of products, particularly those with potentially significant health and safety risks. These registrations include the assessment and periodic audit of the effectiveness of a manufacturer's quality system. Most of these quality-systems registration schemes use the ISO 9000 series standards on quality management and quality systems as the requirements. The ISO 9000 standards and their bases have been adopted in the United States as American National Standards and designated as the ANSI/ASQC Q9000 standard series. As discussed earlier, the joint ANSI/RAB program has been established for accrediting U.S. quality system registrars. This program also contains requirements for quality system auditors.

While there are no mandatory requirements in the United States for quality system registration or approval, some federal agencies are considering how such a step might be useful within their regulatory and procurement programs. For instance, the FDA has replaced its Good Manufacturing Practice (GMP) guidelines with requirements based on ISO 9001. Such a requirement affects any electrical and electronic devices used in medical applications. Other federal agencies are closely studying this issue.

There are compelling reasons why businesses in the United States are becoming keenly interested in the EU developments regarding mutual recognition of quality system certificates and harmonizing certification procedures through European Quality Assessment and Certification Committee (EQS). Third-party assessment and registration of supplier quality systems benefits both the supplier and purchaser. This will occur when the EC, in trading with other countries, permits mutual recognition of registrations. Of course, both supplier and purchaser benefit by expending less resources either auditing or being audited. Suppliers may use their registrations as marketing tools, and results of professional assessments provide suppliers with valuable input for their own quality improvement.

Utilization of both product certification and quality system registration by third parties are becoming a "usual procedure" in several branches of industry around the world. Big markets need more confidence in the quality capability of suppliers, especially foreign suppliers. However, if those registrations are only recognized nationally, it would cause barriers to international trade. It is, therefore, necessary to harmonize procedures to reach mutual recognition of test results and certificates not only within Europe or between Europe and other countries but also internationally. Both the IEC and ISO can play an important role in this area, particularly with their recently established joint program on Quality System Assessment Recognition (QSAR).

LABORATORY ACCREDITATION

Product-related Laboratory Accreditation Programs

Today's emphasis on quality has also heightened awareness in the United States and elsewhere of the importance of good data and competent testing laboratories. Laboratory accreditation was developed as a mechanism for recognizing this competence, yet its use throughout the conformity assessment process varies tremendously. Interest in the accreditation of testing laboratories is increasing in the United States with the focus on special interest areas. The U.S. Congress is passing new legislation each year that creates required accreditation programs. Examples are given in the following text.

> *Fasteners*: The Fastener Quality Act sets up laboratory accreditation as the method to control fastener quality. In this case, the National Institute for Standards and Technology (NIST) must establish its own system to accredit laboratories. Also, it must set up a procedure for recognizing other laboratory accreditation systems operating in the private sector.
>
> *Lead*: EPA has been directed to set up a program to qualify laboratories for testing lead in paint and other environmental media related to the lead remediation efforts. Emphasis will be on use of the private-sector programs.

Pesticides: The Department of Agriculture (USDA) and the FDA have been directed to set up a laboratory accreditation procedure for those laboratories testing for residual pesticides in food. In this case, USDA has decided it best knows how to accomplish this and apparently has decided not to use the private sector.

Asbestos: The U.S. Congress directed that all laboratories that test asbestos in relation to asbestos elimination projects in public schools must be accredited by NIST. This may or may not be extended to other construction sectors in the future.

In the meantime, the federal agencies are designing laboratory accreditation programs under more general provisions of their enabling legislation.

General Laboratory Accreditation Programs

Agriculture: There are at least six programs operating in the USDA. The department's philosophy has been that they are the only ones who know what they are doing so they cannot use the private sector.

FDA Toxicology: FDA originally developed the Good Laboratory Practices (GLP) for use in recognizing the adequacy of long-term feeding studies in the assessment of toxicity of food products. They talk of extending the idea to broader testing areas.

Federal Communications Commission (FCC): FCC has procedures to minimize radiation interference from consumer products. Although it did not view its program as laboratory accreditation, it has those features. FCC is accepting accreditations performed by NIST and the private sector.

Department of Housing and Urban Development (HUD): HUD has for years distinguished between laboratory accreditation and product certification and accepts both NIST and private-sector accreditation. A new program related to lead-based paint is being implemented in cooperation with the Environmental Protection Agency (EPA).

National Voluntary Laboratory Accreditation Program (NVLAP), NIST, Dept. of Commerce: This program was designed for use when a private-sector program is not available to serve the need. In some cases, such as asbestos and fasteners, NIST/NVLAP have been written into the legislation. In other cases, such as insulation and construction materials, the program was established before a private-sector program was clearly available. State and local agencies implement accreditation programs. Some 31 systems in 15 states serve as examples. These systems are primarily in two areas: construction materials and environmental (drinking water).

OSHA: OSHA's Nationally Recognized Testing Laboratory (OSHA/ NRTL) much more directly recognizes product certifiers, many of which are also laboratories.

Private-sector Programs

The private sector has created many separate laboratory accreditation systems with varying types of requirements.

Chemical Manufacturers Association: A program focused on pesticides. Declined to use available systems in favor of establishing their own.

American Welding Society: Accredits welding laboratories at training sites and declined to use existing systems in favor of their own narrowly focused system suitable for their limited purposes.

General Motors, Chrysler, Ford: Used to support "just-in-time" manufacturing processes, where to accept components or subassemblies, the supplier must have their laboratories that supply data showing compliance with specifications accredited. They do their own assessments but are accepting other private-sector systems as well. With so many actors, all with different backgrounds and not all familiar with the vast amount of available information concerning accreditation principles, it is no wonder that confusion has developed!

Worldwide, international, and regional agreements such as the GATT and its successor the World Trade Organization (WTO), NAFTA, and a variety of EU 92 Directives address laboratory accreditation and the acceptance of test data. In the world marketplace, purchasers are stressing quality system compliance with ISO 9000 as a means of ensuring product quality often without a clear understanding of the concepts involved or the relationship among laboratory accreditation, quality system registration, and product certification.

European nations have established the Western Accreditation Cooperation (WELAC). The WELAC create a forum for arriving at a multilateral agreement. The basic structure is that representatives from the laboratory accreditation systems that are members of WELAC perform an assessment of an applicant laboratory accreditation system on behalf of all systems in the agreement. If the basic requirements are met, then the accreditation is recognized by all systems party to the agreement. WELAC has taken the position it would much prefer to establish agreements with parties to other mutual recognition agreements (MRAs) around the world than to provide an assessment team for each applicant accreditation system. The need for recognition agreements will become even more important as the EC works out the practical aspects of recognizing notifying bodies in countries outside the EU.

The electrical manufacturing industry in general supports the accreditation of laboratories to test the conformance of electrical products to standards. But our support is predicated on two conditions: (1) accredited laboratories must utilize and maintain a single test standard for each product and assure consistency in the interpretation of test results, and (2) accreditation systems should be developed in the private sector only when unique circumstances justify such a procedure. Any government-sponsored programs, where necessary, should also implement the concept of a single test standard and assurance of consistency in the testing and interpretation of results.

MANUFACTURER'S PERSPECTIVE OF INTERNATIONAL CONFORMITY ASSESSMENT

In recognition and support of the free enterprise system, and the encouragement of traditional basic supplier-customer relationships, electrical manufacturers believe that ease of market access should be the objective of any international arrangements for conformity assessment. These would cover product testing, certification, and quality-systems registration programs. Product flow across international borders should be assured with a minimum of cost and restrictions.

There should be minimum, if any, governmental involvement in any international conformity-assessment scheme, recognizing that in certain cases where there is no demonstrated voluntary system, there may be need for governmental involvement. Any government regulation should be limited to the focused product sector and should be based upon safety, health, or environmental reasons. The decision-making process should involve the manufacturers of that sector as well as other relevant parties. Conformity-assessment systems should be pursued in a manner that reflects the principles of the Technical Barriers to Trade Agreement of the WTO.

Recognizing the importance of the supplier-customer relationship in pursuing international conformity-assessment schemes, the following are options for conformity assessment:

1. Company reputation (brand-name recognition);
2. Supplier's declaration;
3. Certification established through voluntary or nongovernment arrangements, such as interlaboratory agreements, when marketplace driven (such as the IECEE);
4. Private-sector agreements on conformity assessment activities (''subcontracting''); and
5. Government approval (but only to the extent deemed absolutely necessary).

Except where mandated by law, the use of a third-party test house should be the supplier's choice, which may be made in conjunction with the customer but not by government requirement. Insofar as possible there should be no artificial constraints to influence the dynamics of the marketplace.

AN INTERNATIONAL STRATEGY FOR CONFORMITY ASSESSMENT

The IEC is producing global standards for electrotechnical products at great speed, and its associated bodies, IECQ, IECEx, and IECEE, are attempting to match the demand of sellers and buyers with regard to components and products that meet the relevant international standard. The strategy, therefore, must be to converge the various schemes that already exist and thereby reduce costs and times involved in obtaining conformity assessment.

To achieve global conformity assessment in one step is probably impractical due to the variation in progress achieved in the various regions. Progress will be achieved by seeking agreement between existing international and regional schemes and encouragement should be given to the worldwide use of international schemes. Alternatively, countries should be encouraged to establish acceptable regional conformity-assessment organizations that would simplify procedures by establishing mutual trust. An essential prerequisite of this step is that the international schemes would be accepted to a large extent by the particular country involved.

It is essential that conformity-assessment schemes be efficient and responsive to market demand. Increased up-front initiation costs will be balanced by a reduction in costs associated with the present system of conformity assessment to manufacturers and purchasers. As the boundaries between components and products become even more indistinct, examination should take place to see if international systems can even converge (e.g., IECQ versus IECEE).

Conformity assessment is for the benefit of suppliers and users worldwide but it is essential that those suppliers are aware of its benefits and, therefore, demand and work for their implementation. Successful international conformity-assessment aims, initiatives, and all benefits must be marketed to relevant audiences. In addition to the benefit that the above would provide in monetary terms, there would also be the real, additional benefit to sellers and buyers in the time taken to assess conformity.

In summary, globalization has gained tremendous momentum in recent years. Manufacturers may not have a choice other than meeting head-on the export challenges and they will need to conform to the international or foreign-national testing and certification requirements. The extent of our success in influencing the conformity-assessment process will have a profound effect on the economic future of the countries of the world. It is important that there be a close cooperation between industry and government to establish a coherent and credible conformity-assessment system.

REFERENCES

1. RB Norment. Conformity Assessment and International Trade. Report for the U.S. Consumer Product Safety Commission. Falls Church, VA: Norment and Associates, 1993.
2. International Organization for Standardization. Certification and Related Activities: Assessment and verification of conformity to standards and technical specifications. Geneva, Switzerland: ISO, 1992.

10

National Versus International Standards: Products and Processes

This chapter is contributed by Stephen P. Oksala, Director of Corporate Standards Management for Unisys Corporation. He has also served as as Director of the American National Standards Institute (ANSI). This paper was the winner of the 1996 World Standards Day Paper Competition sponsored by the Standards Engineering Society (SES). It first appeared in *Standards Engineering* 48, no. 6, November/December 1996, the publication of the SES and later republished in *ISO Bulletin* in 1997. It is reprinted with permission of both the author and the SES.

INTRODUCTION

In this chapter I intend to discuss whether national or international standards are preferred. More precisely, I will discuss not only national versus international standards (the products) but also national versus international standardization (the processes). The following paragraphs cover the importance of the subject, the issues in selecting national versus international standards, and the corresponding issues in standardization. In closing, I will offer some specific suggestions for improvement.

The conclusions reached in this essay are straightforward. First, a global economy demands international standards, so there should be minimal use of national standards. Second, there is still a need for a national standards process, albeit with a somewhat different role. Although these two assertions might seem inconsistent, they support each other in the business of facilitating commerce through the use of standards.

WHY CARE ABOUT NATIONAL VERSUS INTERNATIONAL STANDARDS?

The world has always worked on the basis of a mixture of national and international standards. So why should we worry? What has changed to make the use of national or international standards worth examining? For one thing, the marketplace has put more emphasis on standards, so meeting those standards has become more important to suppliers. Industry has also realized the necessity for standards in areas where interoperability between computer systems has become essential for new business methodologies such as electronic commerce. Most importantly, however, public and private authorities rely increasingly on the private-sector voluntary-standards process as a mechanism for generating the technical specifications needed in regulation and procurement.

There have always been close links between standards, regulation, and public procurement. However EC 92, the harmonization of regulation in Europe, made a significant difference. In the New Approach Directives (European-level law) the European Union (EU) made the strategic decision to specify desired results (for example, the product must be safe) and then rely on private-sector standards for the technical details. The directives do not require conformance to standards, but their presumption is that meeting the standards meets the legal requirements. This means that, in practice, the standards become the regulations. The standards process thus becomes critical because once the standards are established and referenced, they cease to be voluntary.

On the procurement side, there is also an emphasis on standards. European public procurers of information technology equipment, for example, generally must choose equipment that meets regional or international standards. This same standardization emphasis carries over to testing and certification since these activities are based on standards for laboratory accreditation and product labeling.

Other countries have begun to follow the European lead. The U.S. Department of Defense (DoD) has embarked on an intensive process to replace its standards with private-sector standards. The Federal Communications Commission (FCC) now allows products to meet standards of the International Electrotechnical Commission (IEC) for electromagnetic interference as an alternative to the commission's own standards. The U.S. Congress has recently incorporated the recommendations in the Office of Management and Budget's (OMB) circular OMB 119 into legislation that directs government agencies to utilize private-sector standards wherever possible. This legislation also authorizes the National

Institute of Standards and Technology (NIST) to coordinate state and local as well as federal use of standards. Other countries are using international standards, their own national standards, or the standards of other nations rather than establishing their own technical bureaucracies.

Finally, nothing has galvanized the public and private sectors into "standards awareness" more than the emergence of management standards. The European New Approach Directives, as noted earlier, specify general requirements. They also specify the methodologies that the manufacturer must use to demonstrate compliance. Furthermore, the New Approach Directives established a new philosophy in this area by introducing the concept of a supplier's declaration of conformance coupled with a third-party verification of the supplier's quality system.

The reliance on the International Organization for Standardization (ISO) 9000 quality-management standards to meet this requirement resulted in dramatic increase in the use of standards, first in Europe and now worldwide. It also led to the creation of an entirely new industry dedicated to certifying companies to those standards. Whether these standards and their use have actually added value is a point of contention, but their commercial and political success is unquestioned.

All standards are more important today. What makes the subject of international versus national standards particularly important is the realization that we live in a global economy. U.S. manufacturers want their businesses to grow, and that means moving into international markets. At the same time, manufacturers outside the United States—some in countries that had no industry in the past—are entering our markets and competing with U.S. firms. Manufacturers in electronics, apparel, information technology, and automobiles, both here and abroad, are competing for customers all over the world. Furthermore, the creation of the World Trade Organization (WTO) reduces the ability of nations to use tariffs, regulations, and nonregulatory "technical barriers to trade."

This focus on global trade makes varying national standards a significant hindrance to growing the world's business. Manufacturers would like to develop a single version of a product, have it tested once (at most), and receive a single certificate of conformance that customers everywhere will accept. While that may not be possible in all cases, I believe it is technically feasible in most. It is, therefore, important that we examine why we still have both national and international standards.

NATIONAL AND INTERNATIONAL STANDARDS

Once we accept the basic premise that free trade on a global basis is always preferable, then there are no good reasons to have national standards. However, if that's true, why do we have so many national standards—and why do we continue to develop them? The most obvious answer is history. Standardization has been a national process for many years because industry has, in the main, stayed within national borders. Only in recent years has the world recognized that global markets are critical to national economies. In addition, all industries have an installed base of products built to unique national standards, so changing those standards is often difficult or impossible. Consider, for example, the obvious difficulty of adopting a new standard for the basic electrical plug. However, if history were the only factor we should see the number of national standards slowly decrease as we focus on international markets and make older technologies obsolete—but we don't. Why?

First of all, there may be a requirement that demands a standard that is useful for one nation but not for all. A desert nation, for example, may have standards for automobiles that mandate higher operating temperatures and greater capability for filtering airborne sand. Such a standard would not be useful to most other nations, and they would probably have no interest in making this an international standard. In fact, they might have a concern that such a standard might force more stringent requirements (and higher costs) on their products even if they did no business in that market.

Unique national standards are most frequently prepared in conjunction with or in response to unique national legislation or regulation. Nations, as represented by their national governments, have different ideas about what needs to be standardized. These differences most often arise because governments deem different kinds or levels of health, safety, or national security measures necessary. Past examples include machine safety and electromagnetic interference, while more recent examples are environmental friendliness and data encryption. In each case, there have been national requirements—leading to a national standard—that differ from the corresponding requirements of other nations.

In some cases these requirements will conflict with each other. A nation that has real concern about fire protection because of the predominance of wooden buildings will develop consumer product standards that require fireproof materials. However, another nation, where concrete buildings are the norm, will be willing to accept less protection in order to reduce the toxicity of the materials in those same products. Even as simple a thing as the color of certain electrical wires in a product may be the subject of conflicting standards.

Whenever these conflicts arise, global companies must produce multiple versions of the same products or modify the basic product. It should be possible for nations to reach agreement on what technical requirements are necessary to meet the needs of the citizenry. However, without hard work, we will continue to see local legislative bodies and regulatory agencies "do their own thing" and thus continue the divergence of national requirements and standards.

Another process leading to divergence is the incorporation of differing national design practices in standards. In many cases, this is due to the practice of developing design standards rather than performance standards. With performance standards, any method can be used as long as the desired results are met. However, if the design is built into the standard, then the national practices will be enshrined in the product requirements. Most developed nations have their favorite design practices, and the national standards process is frequently a vehicle for maintaining those practices in the face of competitive methods. A good example of this is the design of closures for 55 gallon drums. The applicable European standard includes design features common to European manufacturers. Since these design features are not common to U.S. designs, the U.S. manufacturers are at a competitive disadvantage, even though their products meet the requirements as well as the European products do.

Industry can frequently do something about semiunique requirements, regulatory requirements, and national design practices when it sees a global need. However, many industries have not recognized this. The United States has the standards of the ANSI Accredited Standards Committee X12 for electronic data interchange, while the United Nations (UN) oversees the development of the international EDIFACT standards. X12 has tried to adopt the international standards, but those who see no value in global business (small trucking firms, for example) resist the cost of change. Eventually, these firms will see the value of global standards. But, in the meantime, we continue to maintain and enhance national standards.

Unrelated to product requirements, to the needs of customers, or to the needs of producers is the problem of national standards as bureaucratic overhead. Some nations insist that legislation, regulations, or public procurement specifications reference national standards even when the requirements are identical to international standards. For example, the European Committee for Standards (CEN) and the European Committee for Electrotechnical Standards (CENELEC) standards created at the regional level must be adopted without change by each National Standards Organization (NSO) just as each member state must adopt European directives as law. Except for language translation, such a process provides no added value to producer or consumer. However, it does provide benefit to the NSO, which keeps its place in the structure and receives significant revenue from the process. Consider the announcements where NSOs have adopted the ISO 9000 standards. The announcement will merely specify the national identifier, typically some variant with the digits 9 and 0 in it. This process serves no purpose to producers or consumers who will still call it ISO 9000. The NSO played its role, but it provided no value. Everybody else suffered a little bit by having to deal with a more complex (and expensive) process.

While these reasons for national standards are of questionable value, they are frequently a ''cover'' for the real purpose of a national standard—establishing a trade barrier against the products of foreign suppliers. This is most often done by enshrining some arbitrary historical design practice, but, in some cases, a new standard may be created. As an example, a European standard is being created that includes the required dimensions for emergency exits from construction equipment. These dimensions are slightly larger than the standards used by U.S. manufacturers of such equipment. Since the U.S. firms had done extensive research to make sure that they had specified the largest dimensions needed, one can only conclude that disadvantaging U.S. manufacturers was a primary consideration in the standards effort. This process has not been uncommon over the years in standardization. Nations and the people and organizations within them have a strong desire to protect their own because of an "us or them" mentality or because local industry is not up to competing with foreign firms.

In summary, national standards arise for a variety of reasons, none of which are positive in a global economy. Industry should strive for international standards yet recognize that we have organized our world into nations and that national loyalty is one of the most powerful forces in human society. Industry needs to push for change, but carefully and with the recognition that change will take significant time to accomplish.

This discussion has not addressed the phenomenon of regional standards, although organizations such as the EU, the North American Free Trade Agreement (NAFTA), and the Asia Pacific Economic Cooperation (APEC) are growing in popularity. In my opinion, standards arising from such groups suffer from the negative aspects of national standards and may introduce another level of approval into the process. On the other hand, they can reduce the total number of different standards. They are an intermediate process that is likely to continue because of political considerations, and we need to deal with them in much the same way as we deal with national standards.

NATIONAL AND INTERNATIONAL STANDARDIZATION

While single international standards are clearly desirable, the discussion about national versus international standards processes is a little more complex. The previous paragraphs discussed international standards in terms of their acceptance, without being specific about their origin. But how they came to be is precisely the point in a discussion about standardization, the process of developing standards. When the question "What is an international standard?" is asked, the questioner usually wants something more definitive than "a specification that is accepted and endorsed on a world-wide basis." Users of standards want something that can make distinctions based on the organizations that create them because that will give some assurance that spending resources in that venue is most likely to provide results. Let's say that an international standard is any standard produced by an international standards organization. But what is that?

Ever since the modern era of standardization began at the end of the last century, the world has defined international standards organizations (like their counterparts in governance) as those which have nations as members. Administrations (governments) are the members in the International Telecommunication Union (ITU), and there are National Bodies in ISO and National Committees in the IEC. An official (designated by the government) NSO typically represents the country in the latter two. In the United States, ANSI is the official representative to ISO; the U.S. National Committee (USNC), a subgroup of ANSI, is the representative to the IEC; and the U.S. Department of State handles ITU. Most other nations have similar arrangements. At the international committee and subcommittee levels, a single national delegation with a single point of view represents each nation. At the working group level of ISO and IEC, where participants are technical experts expressing their own views, the NSO must still nominate those participants.

Today there are significant attempts to change this paradigm. Some standardization organizations, which in the past were content to be part of a national process, consider themselves to be international or "transnational" because their membership comes from many countries. Some, such as the Institute for Electrical and Electronic Engineers (IEEE), are accredited by ANSI and work through the formal U.S. standards process. Others, such as the Internet Engineering Task Force (IETF) or the ATM Forum, do not. There are also industry consortia that produce technical specifications under widely varying participation and consensus requirement rules.

The problem today is that the standardization products of the organizations working outside the formal system are being widely adopted, often in preference to the products of the formal bodies. People perceive that the new groups are able to produce standards more quickly. This is based on a number of factors, but the ones relevant to this discussion are (1) the absence of a requirement to involve all interested or affected parties and (2) the elimination of the need to develop a single national position. The latter point means that multinational companies can advocate their views directly rather than going through national processes that may change or even ignore their positions.

The information technology industry has expressed the most concern about this problem. In response, ISO/IEC Joint Technical Committee 1 For Information Technology (JTC 1) has adopted procedures to allow such organizations to bring their "publicly available specifications" directly into the international process. ISO and IEC are considering expanding this kind of relationship. If international standards are the desired end point, is this direct recognition of other organizations the beginning of the end for national standards processes?

It is my belief that there are and will continue to be good reasons for having national standards organizations and national standards processes. Perhaps the most significant is the point mentioned earlier—the world works through nations, and technical experts cannot successfully ignore the effect of standards on their governments or on their national economies. Beyond that general observation, there are a number of roles that an NSO can play, even if other organizations take on the primary role in developing technical specifications.

Representing All the Voices

Even with the most advanced electronic methods, the standardization process will continue to involve periodic meetings in which significant work will take place. Many will be unable to participate directly in such international assemblies because of cost constraints. Academics, small and medium enterprises, and non-profit groups such as consumer advocacy organizations are less likely to attend, even though they have interest and are materially affected. Likewise, government agencies are unlikely to be able to participate directly in an international process. We probably do not want multiple agencies participating (consider the possibility of the FCC arguing with the Commerce Department in an ISO meeting) or power-ful government representatives whose views might drown out other participants. If we really want general international consensus to back up an international standard, and I believe we do, then there needs to be a mechanism through which the process can hear those voices. A national standards organization working to address the topics in the national context and for national interests can provide necessary input even if the major players can ''go direct.'' That same NSO can be a voice in keeping the number of international organizations at a manageable level (meaning as few as possible) through its influence in both the public and private sectors.

Helping National Industry Compete

There is a need on occasion to provide a vehicle by which a national industry can win a standards battle. We like to think of standardizers uniting in the interests of international harmony. All too often, however, national private- and public-sector participants want to standardize a particular solution to benefit their local industries. A national standards organization can focus these efforts and make it more likely that the group will succeed. Is this competitive activity a plus for the standardization process? It usually depends on whether you win or lose. How-ever, it is a necessary tool in a world where national considerations really do make a difference.

Information Pipeline and Service Provider

With all the organizations involved in standardization, there is definite value in providing information to companies and individuals on what standards exist and where to get them. These are also services that a recognized national body can provide to people who are truly confused by the Byzantine structures of national and global standards processes. NSOs can train participants, assist groups in utilizing new technology, and advise on how to provide due process and avoid legal pitfalls. The NSO can also manage the relationships between standardizers, testers, and certifiers and even assist in funding start-up activities. All of these functions have the potential to enhance the effectiveness of individual standards developing groups.

Representing Industry to Government

There are times when industry needs to speak ''as one'' on its interests to the government. This has to be a national process, and there is obvious value in an organization that can deal with the government on testing and certification, public procurement, and other standards-related issues. The NSO is the only organization in a good position to do this.

Maintaining the Current Process

Even in the new world of international commerce and electronic communications, there will always be situations where the participants cannot or do not wish to establish a direct international process. Therefore, there is still a need for a national process to provide the vehicle through which interested parties can participate in international standards development. Not all industries have the same needs; it is important for a national standards process to be flexible in how it accomplishes its objectives.

Finally, it is worth noting some things that perhaps should no longer be the responsibility of the national standards process as embodied in classic NSOs. They have historically acted as gatekeepers, limiting participation to eligible parties. Instead, they could consider their major role to be facilitators in the process of national standards development, emphasizing new ways to involve all the interested parties and providing assistance in dealing with the outside world. With the proliferation of consortia, fora, and similar organizations, perhaps the focus should be less on accreditation (a gatekeeping function) and more on publicity and communications. We also need to make sure that the national process is not an end in itself. We need to take a hard look at NSO businesses and consider alternate ways of financing the process.

SPECIFIC SUGGESTIONS

The preceding sections of this essay suggest that more specific actions exist that will improve the standards system. So in conclusion, I would like to offer some ideas for discussion. The standardization community may not be ready or able to implement them today, but they can serve to spark debate on how to improve the standardization system.

Encourage NSOs to establish a policy whereby formal international standards are automatically adopted as national references (a simple, one-page reference to the international standard) unless there is significant objection from affected parties. This would maintain some valid national control, while eliminating the time and effort required in 90% of the cases that require no real action.

Establish a policy that anyone can participate in the international technical process without needing anyone else's permission. At the approval level, I believe that the national balloting process should be retained to ensure that there is real consensus. However, in preparing the initial specification, we should welcome the contributions of everyone.

Allow private organizations to be direct participants at the technical committee level and possibly even at the top administrative level. Consider whether they should have the right to vote. They may represent more people or economic power than some nations!

Establish a policy and process that mandates written justification for proposals for strictly national standards. Consider establishing a new category of ''candidate international standards'' for work to be taken to the international world when there is enough international interest or when the national group has developed a national industry consensus.

Simple? Not necessarily. Practical? Maybe—at some point. Certainly, we need to rethink the roles of the national, transnational, and international organizations in a global economy. These suggestions may be a place to start.

11

Introduction to ISO 9000 and ISO 14000 Management System Standards

Dr. Steven Spivak, coauthor of the book, is Professor and Chair, Department of Fire Protection Engineering, The University of Maryland. From 1991–1995 he served as chairman of the International Organization for Standardization (ISO) Consumer Policy Committee (COPOLCO), one of only four ISO policy committees and which reports directly to the full ISO General Assembly. Prior to that he served for many years as member and head of the USA/ANSI delegation to ISO/COPOLCO. He is presently serving a fourth term as a Director of the American National Standards Institute (ANSI) (and U.S. member body to ISO and International Electrotechnical Commission, ICE), and is a member of the Corporation of Underwriters Laboratories (UL).

OVERVIEW

This chapter serves as an introduction to David Jones' invited chapter "ISO 9000 and the Real World" which follows. This overview gives a brief description of the worldwide importance of ISO 9000 international quality management system (QMS) standards, and the growing significance of the ISO 14000 international environmental management system (EMS) standards (currently ISO 14001). In addition, many iterations and variations of management system standards are flourishing throughout various sectors of business and industry. From a humble beginning among the quality standards and their precursors in the 1960s and 1970s, ISO 9000 and third-party or independent registrations (known in Europe as *certifications*) now probably number over 50,000 plant sites and/or business

101

entities in North America and about 200,000 worldwide. Even the widely read and syndicated comic strip *Dilbert* is replete with ISO 9000 cartoons and frustrated business managers having to create anew, cope with, or curse about these same quality standards. From the sometimes arcane world of technical standardizers to having now become the common vocabulary of business and government leaders, one must now conclude that voluntary international standards such as ISO 9000 and now ISO 14000 have "arrived" on the business scene. Paraphrasing another popular advertisement of the 1990s, "You've come a long way, ISO."

Much of the relevant experience for ISO 9000 can be traced to the British Standards BS 5750 of 1979, and other related quality standards in several nations such as the 1959 military standard MIL-Q-9858 of the U.S. Department of Defense (DoD); the related 1968 AQAP quality-assurance program of the North Atlantic Treaty Organization (NATO); Canadian standard CSA 7299 of 1975; and many others. Recent variations include the QS-9000 QMS standards of the "big three" United States–based automobile manufacturers, and in particular their materials and components suppliers; AS-9000 being implemented in the aerospace industry; and others. The following experiential chapter is a recounting by David Jones of Robinson Bros., United Kingdom and the experiences of a quality-assurance manager who has continued to guide their company through the processes of quality, environmental, health, and safety management.

The ISO 9000:2000 series consists of three (formerly five) core standards or guidance documents, ISO 9000 to 9004. Several other related quality standards, terminology standard for quality, certain ISO 10000 series standards on quality auditing, and more can be found in the comprehensive collection, *ISO Compendium* of quality related standards.

The five basic quality documents in the ISO 9000:2000 series are:

ISO 9000: "Quality Management Systems—Fundamentals and Vocabulary"
ISO 9001: "Quality Systems—Requirements"
ISO 9002: "Quality Systems—Model for Quality Assurance in Production and Installation" (now included in ISO 9001:2000).
ISO 9003: "Quality Systems—Model for Quality Assurance in Final Inspection and Test" (now included in ISO 9001:2000).
ISO 9004: "Quality Management Systems Guidance for Performance Improvement"

These standards spell out the requirements for establishing quality-assurance programs within a company. The original efforts within ISO to "internationalize" new quality management system standards originated within ISO Technical Committee (TC) 176 with participants who were of national associations and ISO members Association Francaise de Normalisation (AFNOR), ANSI, British Standards Institution (BSI), Netherlands Normalization Institute (NNI), and Standards Council of Canada (SCC). Other member countries were also represented and so it became truly a global effort. Today almost every country in the trading world has either adopted the ISO 9000 and some related ISO 10000 quality standards, or has its own identical or almost identical national iteration.

These ISO standards may be referred to by different notations in each country's national standards system, where identical versions of the ISO series appear. For example, in the United States they would be ANSI/ASQ 9000, 9001, 9002, 9003, and 9004. The most recent editions are being published during 1999–2000 and future versions as draft international standards (ISO DIS) may become available. The ISO 9000 and 9004 standards of the series are guidelines supporting the other three documents. They help users decide which model to use for quality improvement and lead to possible quality registration. None of these documents specifically set forward minimum standards such as strength, conductivity, color, density, etc., for products or for services. The standards ISO 9001 to 9003 ''. . . complement relevant product or service requirements given in the technical specifications'' (ANSI/ASQ 9000).

The purposes of these standards are to ensure that the seller's operational procedures and management practices will produce the product quality expected by the customer. It is vitally important to recognize that the ISO 9000 compliance and independent audit, with concomitant registration or certification for successful compliance, applies only to the quality-management system of the company. Although registration or certification has real value (as Chapter 12 will document), it does not imply nor mean that the actual product or service of the registered company is ISO registered, or approved, or of inestimable quality, or ISO 9000 registered, per se. This is a common confusion, or even misrepresentation, on the part of some neophytes to management systems standardization, and this distinction must be noted.

The conformity assessment system and standards for EMS have been written by ISO TC 207 and have resulted in the ISO 14000 series of standards. Currently published or in draft form are the following:

ISO/CD 14000: ''Guide to Environmental Management Principles, Systems and Supporting Techniques''
ISO 14001: ''Environmental Management Systems—Specification''
ISO/CD 14010: ''Guidelines for Environmental Auditing—Audit Proceures Part 1: Auditing of Environmental Management Systems''
ISO/CD 14012: ''Guidelines or Environmental Auditing—Qualification Requirements for Environmental Auditors''
ISO/CD 14031: ''Generic Environmental Performance Evaluation Methodology''

This brief discussion is sufficient only to provide a ''setting'' for Chapter 12. Any corporation or organization planning to implement these standards is well advised to seek out the voluminous references and resources available on ISO 9000, ISO 14000, and other management system standards. There are numerous books that describe in detail the methods to achieve certification, and especially those companion titles published by Marcel Dekker. We have listed in the reference section a few such works and Internet Web Sites related to quality or environmental standards. A search of the Internet will reveal more.

REFERENCES

1. JL Lamprecht. ISO 9000—Preparing for Registration. Quality and Reliability/32. New York: Marcel Dekker, 1992.
2. JL Lamprecht. Implementing the ISO 9000 Series. Quality and Reliability/40. New York: Marcel Dekker, 1993.
3. G Naroola, R MacConnel. How to Achieve ISO 9000 Registration Economically and Efficiently. Quality and Reliability/48. New York: Marcel Dekker, 1996.
4. M Breitenberg. Questions and Answers on Quality, the ISO 9000 Standard Series, Quality System Registration, and Related Issues. NISTIR 4721. More Questions and Answers on the ISO 9000 Standard Series and Related Issues. NISTIR 5122. Gaithersburg, MD: National Institute of Standards and Technology, U.S. Dept. of Commerce, April 1993. (Available from National Technical Information Service, Springfield, VA.)
5. International Organization for Standardization, ISO 9000 News, Geneva, Switzerland: International Organization for Standardization. ISO Compendium, Implementing ISO 9000 descriptive brochure; and several other ISO 9000 or 14000 materials available from ISO or from a national standards body (ANSI—USA, SCC—Canada, DGN—Mexico, commercial vendors, and others as appropriate).
6. Internet quality-related Web sites include, but are not limited to, the following: http://www.quality.org/; for ISO see http://www.iso.ch; for ANSI see http://www.ansi.org; for NIST/US Department of Commerce see http://www.nist.gov and http://www.quality.nist.gov; for ISO 14000 explore http://www.iso14000.org; http://www.mgmnt14k.com/; or http://wineasy.se/qmp; and for Quality Digest see http://www.qualitydigest.com
7. International Organization for Standardization. Publicizing your ISO 9000 or ISO 14000 certification. Geneva, Switzerland: ISO Central Secretariat, 1997. Guidelines for proper advertising and promotion of ISO-based registration or certification.
8. SC Puri. Stepping Up to ISO 14000: Integrating Environmental Quality with ISO 9000 and TQM. Portland, Oregon: Productivity Press, 1996.
9. D Hutchinson. Safety, Health and Environmental Quality Systems Management, Lanchester Press, 1997. (Available from the American Society of Safety Engineers [ASSE].)
10. S Adams. The Dilbert Principle. New York: Harper Business, 1996, pp 240–243.

12

ISO 9000 and the Real World

This chapter was especially prepared and contributed by David E. Jones, Quality Assurance Manager, Robinson Brothers Limited, West Bromwich, England, United Kingdom and a highly regarded manufacturer of specialty chemicals. The chapter describes their experiences in taking plant sites through successful ISO 9000 audit and quality registration, and added perspectives on environmental, health,and safety standards and management systems.

INTRODUCTION

Robinson Brothers Limited is a medium-sized privately owned chemical manufacturing firm based in Birmingham, England. The company achieved registration to what was the predecessor to ISO 9000, BS 5750 (Part 2), in 1990. The preparation for the registration assessment and subsequently "living with the standard" gave rise to a set of experiences that will be of interest to both standard developers and managers of organizations seeking to implement a formalized quality-management system. In this chapter, I shall share those experiences in the hope that it will assist those who are actively involved in the Quality-Assurance System development to the ISO 9000 standard. I am aware of a debate in the United States about the pros and cons of ISO 9000 certification. Having lived through a similar debate in the United Kingdom, and implemented the standard in practice, my view is that there are more pros than cons. Furthermore, an extension of management system standardization to formalize environmental and health and safety management systems is seen as a natural development receiving support in the United Kingdom, not least from the regulating authorities. The experience of Robinson Brothers Limited in this field will also be discussed.

ISO 9000 (FORMERLY BS 5750)

ISO 9000 is a voluntary international standard that describes a model for quality assurance in the production, supply (delivery), and servicing of either manufactured goods or services. Three models exist to represent differing levels of quality system requirements.

1. ISO 9001 (BS 5750 Part 1)—used by suppliers of goods or services to assure conformance to specified requirements during the design, development, production, supply, and servicing of their products.
2. ISO 9002 (BS 5750 Part 2)—assures that suppliers can produce, supply, and service their products to specified requirements without the design and development aspects (in ISO 9001) having to be formalized.
3. ISO 9003 (BS 5750 Part 3)—offers a model for the assurance of only final inspection and testing (quality control), and is less often used as it finds few valid applications in industry that should not be covered by one of the other two system models referenced previously.

Both ISO 9001 and ISO 9002 require as basic elements documented procedures to formalize and control many activities. By way of example, certain key ones for manufacturing are:

1. Sales order processing (contract review);
2. Raw material acquisition and inspection;
3. Process description and control;
4. Quality control;
5. Control of nonconforming material;
6. Material storage and handling;
7. Internal auditing of the documented quality system; and
8. Training.

WHY SEEK ISO 9000 REGISTRATION?

As quality manager of a chemical manufacturer supplying a range of industries from rubber to pharmaceuticals, I became aware of a growing number of quality questionnaires hitting my desk. These invariably came with a cover letter stating that suppliers who were not in possession of an ISO 9000 certificate may find themselves barred from the approved supplier list. Completing the questionnaires was in itself a tedious exercise, especially as we did not possess the certificate and we would not skip questions 5–25 of the questionnaire.

By 1987, some rationale for certification to ISO 9000 began to emerge. It became part of a drive by the British government to improve the quality status of the British manufacturing industry. However, the distinct engineering bias made interpretation difficult and opened the door to waves of consultants to act as interpreters to us humble chemists.

Total Quality Management (TQM) and, at the time to a lesser extent, Continuous Quality Improvement (CQI) were the way forward. In the United Kingdom, it was presented not as a competitive issue but one of business survival. Having seen the decimation of the U.K. motorcycle industry, and being actively aware of the flood of high-quality goods hitting the shops, the chemical industry viewed their own survival as a key issue, and the consultants were invited in.

Once a few companies in our industry sector gained registration, the majority of them being subsidiaries of high-profile majors, it spurred the rest of us on lest we did indeed become barred from our customers' approved supplier listings. Nobody knew how real that threat was but no one was prepared to test it out. As we will see later, it is our experience that there is indeed objective evidence of improved supplier performance among those who have achieved ISO 9000 registration. In hindsight we think we did the right thing by achieving registration, although I have to confess that I haven't barred anyone from our own approved supplier list for not having ISO 9000 registration. However, new suppliers that are registered are favored over those that are not.

A bandwagon had been started and was further fueled by blanket coverage in journals and on the conference circuit. It (BS 5750) even became a topic of dinner conversation, much to the dismay of the partners of quality managers. To the more astute observer of business matters, it also became clear that organizations of differing sizes had differing agendas. This could be rationalized as the following views of the small, medium, and large U.K. organizations.

1. Small (0–200 employees)—no can do, only for the big boys, another management fad
2. Medium (200–2,000)—a survival issue
3. Large (2,000 +)—a stepping stone to TQM

Those in the medium sector, like ourselves with 420 employees, saw the linkage to TQM and were keen to pursue continual improvement, once we had guaranteed our survival. However, there was to be plenty of trial and tribulation before we achieved registration and then sought to maintain it.

PROBLEMS ENCOUNTERED

Developing the System

As stated earlier, the engineering bias to the standard meant that the pioneers of registration outside of the engineering industry, who themselves had been operating to BS 5750 since the early 1980s, were exposed to consultants from that background. In our experience it meant that "off the shelf" standard systems were deployed that in hindsight could have been much better designed, and most definitely made more user friendly. The consultants presented ISO 9000 almost as a deity. The 18 or 20 paragraph requirements became like the 10 commandments:

Thou shalt have (our) Quality Manual;
Thou shalt calibrate (all) thy equipment; and
Thou shalt send your suppliers (our sample) vendor questionnaires. . . .
(The words in parentheses being the consultants' license.)

Surely there was room for a degree of pragmatism here. Our industry representatives the Chemical Industries Association (CIA) offered some help with guidance on the 1987 version of the BS 5750 standard. It came just in time for us to ward off the excesses of the consultant's approach as we set about our quality-assurance system development. This was in early 1989 after one false start already. The key issues learned were:

1. Don't underestimate the size of the task—especially when starting from nothing in the way of formalized (documented) procedures;
2. Seek help and guidance from within your industry group; and
3. Remember that you have to "run" with this system long after the consultant has gone.

To expand a little on our main problem area, the lack of formalized procedures; the reason we never had any became glaringly obvious. That was because no one, myself included, had ever written one and, therefore, it was a skill that was totally lacking. The outcome was an unfortunate legacy of that lack of in-house expertise, in other words, procedures that recorded compliance with the standard's requirements but did not describe accurately how we actually got there.

As the preregistration assessment approached in 1990, about 18 months into the systems development program, approximately one-third of the top-level procedures were still outstanding. Project management now drifted into crisis management as the assessment day loomed. Many late nights and a concerted effort by a relatively small number of people enabled us to get the last procedure written and issued just one week before the assessment.

Our consultant was distinctly uneasy about this, as his original plan was to run the system for two to three months prior to the registration assessment in order to get sufficient evidence/data for the external assessors to pass judgement on. Luckily the assessors proved to be pragmatists, too. Their focus was on our core activities, namely the manufacture of chemicals from controlled raw materials to satisfy a customer's requirements in full. This we were able to demonstrate was well controlled by our documented quality assurance (QA) system. Of course they had to assess all 18 clauses of the standard over five days, spent over two sites; but only limited coverage of the peripheral activities such as training, statistical techniques, and purchaser-supplied material was offered. A number of nonconformities (even if relatively minor) had to be addressed by suitable corrective actions and we were finally awarded accredited certification to BS 5750 Part 2: 1987 in July of 1990.

We got it and wanted to keep it. One significant feature of the route to registration was the genuine interest and commitment of the vast majority of the workforce, particularly those at the operative level. They, for once, could recognize a corporate goal and did their best to ensure they played their part in helping us achieve it. This was something to build on during the next journey towards TQM.

Maintaining the System

The first thing that slips is the internal audits. This was the voice of an experienced practitioner of ISO 9000 Quality Systems, the comment being made at a training course for lead assessors in quality-assurance system auditing.

In my view, the requirement for internal auditing is the most important clause in the standard. During our first round of internal audits, post registration, we had in some cases our first opportunity to get the real process owners involved in the ISO 9000 system. The quality-assurance manager is responsible for maintaining the system as a whole and for ensuring that continued compliance with the organization's declared procedures can be demonstrated. It was a rude awakening for the line management when it had to be explained to them that the quality-assurance manager—although he may have written the procedure—was not actually responsible for purchasing control, work instructions preparation, etc., on a day-to-day basis.

On our first round of procedure reviews we were able to get to the real story. Second issues were promptly produced, which by and large now reflected what actually did happen in the respective departments and functions. These revised procedures now did read like living documents and not just a report of what was supposed to happen in theory in order to comply with the standard.

Auditing

During 1990 and 1991, the first full 12 months post registration, the external auditors performing the continuing assessment reviews were again a throwback to the engineering sector. It has to be said that they were of poor quality. In a situation where industry knowledge was a distinct advantage in order to provide a positive assessment of opportunities for system improvements, the external auditors stayed in their own comfort zone—usually calibration. Calibration was being done to death. Yes, it is important, but so are many other clauses of the standard.

Within discussions between quality-assurance professionals, inconsistency in external auditor performance was seen as a common problem. The assessment authorities reacted to this criticism fairly quickly, but only just in time to maintain the credibility of the standard that was in danger of being viewed as a calibration standard. A genuine interest by the external auditors in the substance of the management review meetings, and the effectiveness of the internal auditing process was now being shown. That is to say, the wider effects of ISO 9000 certification within the organization were being reviewed.

Four years into certification we began to question the need for detailed internal audits that focused solely on procedural compliance. Not least, the need for some change was equally important to maintain the interest of a flagging internal audit team. Therefore, to be consistent with a revised quality policy, which placed a strong emphasis on continual improvement, we took the view that, save for specific aspects of the system worthy of a full-compliance check on a needs-must basis, we should empower our audit team to identify and record scope for functional improvements. By keeping the internal audit team armed with data on the corporate quality performance, the result was a healthy number of good ideas to improve our system's performance. By bouncing these ideas off the process owners, lasting improvements to the system were seen that also impacted positively on our quality-performance measures, and in particular on-time delivery performance.

The role of the internal auditors needs to be developed over time. In our experience it is beneficial to draw the internal auditing resource from areas other than just quality-assurance professionals. It is also important, therefore, not to overestimate the auditors' knowledge of the quality system. A minimum of two to three years internal auditing experience may be required to at least be exposed to every aspect of the standard. Overall, the quality of our internal audits is good and training is essential in not just the standard but also in auditing skills in order to achieve this quality of audit.

We select our internal auditors as people who we think will be committed to, and then benefit from, self-development. It is seen as a good form of management training. By expanding the auditing brief into both system and business improvement aspects, the quality-assurance system auditors have developed into a corporate auditing team bringing their skills of system auditing to reviews of wider aspects of the organization's activities. This applies particularly in health, safety, and environmental management systems, about which more will be said later.

A Paperwork Culture

The most cited criticism of ISO 9000 is that it is too bureaucratic. I would have to agree with this sentiment. However, to put the standard into context there are many more requirements (shalls) within the text of ISO 9000 (140, so I am told) compared to the nominal 20 main clauses. Each will have to be described by a written procedure, itself a standard requirement. It was therefore, with some interest that I learned about several organizations in the United Kingdom that had initially undertook a TQM initiative and then considered ISO 9000 registration some years into their TQM program. They were actually considering flowcharting their ISO 9000 procedures, most probably as a result of experiences in process mapping, to aid identification of internal customer and supplier interfaces. We are now flowcharting some procedures ourselves, particularly when it is thought that a picture does say a thousand words, or when we wish to clarify a complex process.

Detailed working procedures are a definite turn off for the guys at the sharp end. They just never get read. A skillful pro forma designer is an asset as the recording of key quality-system data or instructions on single sheets of paper is highly desirable. But what about all those procedures anyway, how can they be managed?

Our quality-assurance system has 31 main procedures, the quality manual, and even more procedures that are under development to assist an integrated management-system approach. Some of the quality-assurance procedures have stood the test of time and remain at that Issue 2! Others have been so refined (and improved) that we are on Issue 7, (i.e., the 6th complete rewrite). This is a positive aspect and is indeed a true measure of system improvement. It also demonstrates that an active audit and review process is in operation.

The association between ISO 9000 registration and filling in paperwork was a problem in the early days. People were beginning to wonder if the world had turned turtle and paper shuffling had been delegated to the blue/gray-collar ranks. This certainly was not what empowerment was supposed to be about. In our work with pharmaceutical companies, we have seen the benefit of detailed record keeping to assist in problem solving as well as liability defense. But there is a more-ready acceptance of electronic data storage for ISO 9000 records than for Good Manufacturing Practice (GMP) and it should, therefore, drive us to information technology solutions to record keeping or information transfer problems within the quality system. In the United Kingdom, software for quality-assurance system control, surprisingly, has emerged only recently. This perhaps confirms the fact that effective compliance is best achieved in hard-copy systems. The paperless office will be slow to arrive at the quality-assurance manager's door. The day when the external auditor conducts his continuing assessment down a telephone wire is also some way off.

In concluding this section, it must be reiterated that resistance to bureaucracy may damage general acceptance of the ISO 9000 ideal among an organization's workforce. For those starting off on the road to ISO 9000 certification, a strategy to deal with this problem surely must be an integral part of the project. The acronym "KISS" is often used in this regard, "Keep It Simple, Stupid."

THE BENEFITS OF ISO 9000 REGISTRATION

As already mentioned, even given a certain amount of bureaucracy, the power of effective traceability in a manufacturing chain is a positive outcome of ISO 9000. As an example from our own system, it was with much pleasure (and surprise) that we were able on one occasion to back track to how exactly we did make that chemical four years ago on the other site, and on which of two plants. A genuine inquiry successfully answered. However, the main benefits go way beyond having an effective corporate memory.

We have researched quality costing quite extensively at Robinson Brothers Limited using the BS 6143 model. The Prevention, Appraisal, and Failure (PAF) cost model was adopted with some evidence of the classical relationship whereby an increased expenditure on prevention (QA) gives rise to a decreased failure cost (Fig. 12.1). In some years, the overall reduction in total costs amounted to an increase in corporate profitability of circa 10%.

Our original analysis of this situation was that simply by formalizing all the production operation work instructions, a much more consistent manufacturing operation resulted, thus leading to a significant improvement in first-time quality measures. This was, however, offset in later years (e.g., 1991 and 1995) by the increased transparency of failings in certain manufacturing operations resulting in spikes of increased internal failure costs. On the external failure cost side we are now seeing the costs of failing to meet increasing consumer expectations. This has been particularly prevalent in the rubber industry as the automotive supply chain demands continue to stiffen. Then, from 1996 onwards, as our business focused more and more on fine chemical operations, our reaction to the tighter material specifications was reflected in more stringent quality control (appraisal) with consequential increases in internal failure costs. These internal failure costs need more-focused management intervention to identify lasting solutions to recurring problems. To tackle this issue we now need to incorporate more formalized problem-solving techniques into our ISO 9000 system.

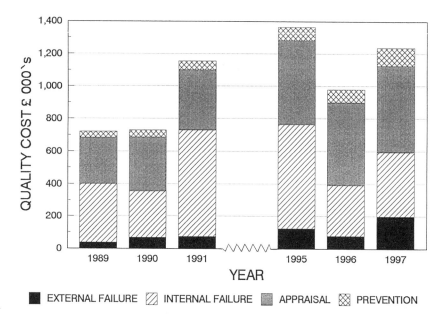

Figure 12.1   Comparison of appraisal, failure and prevention elements adapted to quality costing review.

Supplier nonconformity levels were quite high in our first year of ISO 9000 certification. Our supplier-assessment procedure provides us with a means of assessing the ISO 9000 status of our supplier base.

Figure 12.2 shows that as ISO 9000 attainment increased initially among the supplier base, the supplier nonconformity levels reduced. This was even after experiencing a modest increase in operating activity. The trend reversed in 1994 when a new phase of supplier assessments was introduced and our expectations of quality of service and supply began to mirror those of our customers. This cycle of ever more-stringent supplier controls was ramped up again in 1997 and was reflected by an increase in recorded supplier nonconformities. The underlying message was that procuring from an ISO 9000 certified source did give weight to the argument that this should be a preferred option. Provided that further improvements in supply quality and service are fixed into procurement specifications, then the ISO 9000 system should give some assurance that the continual improvement process is not hindered by supplier noncompliance.

One further benefit of ISO 9000 has been to assist understanding between functions of the organization, as a consistent terminology is used. One example of a high degree of corporate understanding is in the area of the relevant ISO 9000 standard relating to contract review. Personnel in sales, production, quality

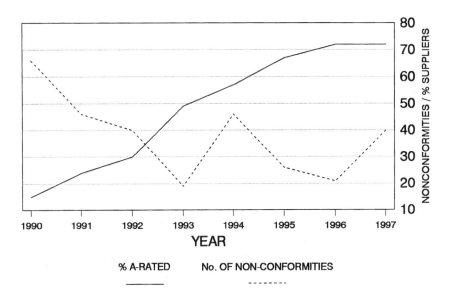

Figure 12.2   Comparison of supplier nonconformity levels with percentage of suppliers fully approved (A-rated).

control, research, and development are now aware of the need to establish and verify that customer requirements can be met in full. The ISO 9000 system will ensure that all the necessary information is made available to the appropriate functions. This way, customer requirements are assured of being met. Prior to ISO 9000 this was not always possible. However, exceeding customer expectations is not necessarily an outcome of an ISO 9000 operation, as the quality cost analysis in Figure 1 illustrates. There is a tendency to only deliver the standard service: only the more enlightened personnel go beyond the norm to ensure total customer satisfaction.

By far, the main benefit of ISO 9000 certification has been to provide a basis for continual improvement. At Robinson Brothers Limited we adopt the ISO 9000 wedge principle to underpin continual improvement—as represented in Figure 12.3. The principle is communicated to all employees. We demonstrate that improved output or service quality may require a change in the way we operate. To ensure that change is fixed within the operating system of the organization the appropriate procedure is amended—or even a new one written. They will then become part of the ISO 9000 system and, therefore, will be subject to internal audit and review. The audit and review process will give us the opportunity to maintain the improvement as we do in the rest of the quality-management system.

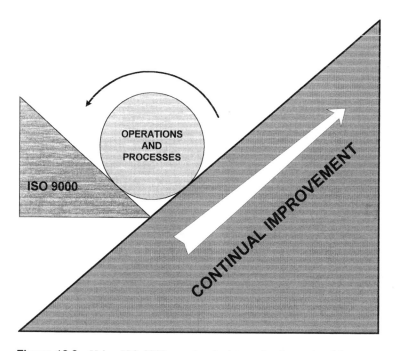

Figure 12.3   Using ISO 9000 as a 'wedge' to underpin continual improvement.

We have undertaken a benchmarking initiative with the intention of identifying best practice in several of our key business areas. Again, any need for changed procedures will be accommodated by, and therefore fixed into, the ISO 9000 system either as minor changes or perhaps even to reflect totally reengineered processes. One final comment on the way that ISO 9000 supports continual improvement is in the application of statistical techniques. The strengthening of this clause in the 1994 revision of the standard was long overdue. Outside of the automotive supply chain there has been little application of this powerful management tool. Making sure that our procedures reference the application of statistical techniques, to nonmanufacturing areas as well, will facilitate the continued corporate understanding of the total quality system and its purpose.

THE FUTURE ROLE OF STANDARDS IN INDUSTRY

In the United Kingdom for the past several years there has been a significant increase in regulatory requirements affecting the manufacturing industry. The regulatory burden of the health, safety, and environmental aspects of our business have, in particular, provided stiff management and technical challenges. The regulatory authority controlling health and safety has declared openly that in order to successfully control risk there must be an effective management system. The pollution inspectorate similarly judge that effective compliance with consented environmental emissions and demonstrable environmental performance improvement can only be maintained by having an appropriate management system.

It is not by chance that parallels with the ISO 9000 system for quality assurance are drawn. However, given the negative comments about the predecessor BS 5750 relating to its bureaucracy, the respective Environmental and Health and Safety Management System (EMS) standards are numbered BS 7750 and BS 8750 and subsequently are tended to be viewed with some cynicism as to their real worth.

Lessons learned from BS 5750 will ensure that the mistakes made during development of that system should not be repeated in these new areas. The pragmatic view must be that management system standards are here to stay. The balance of opinion regarding ISO 9000 is that it has had a beneficial impact on manufacturing quality, but perhaps less so on the output quality-of-service organizations applying that standard. The result is that the medium-to-large organizations can again muster the resource to develop these new management systems. The smaller organizations again are tending to cry wolf. However, whereas you don't have to manage quality, and perhaps can get by environmentally, there is no excuse for not managing safety.

The difference between this new set of standards and ISO 9000 for quality is that business survival is not (immediately) at risk. So what is the real motivation to develop these new systems when third-party accreditation (certification) is not such an important goal?

Environmental Management Systems BS 7750 and ISO 14000

ISO 14001 is, by and large, judged to be totally acceptable to the U.K. chemical industry to describe how a management system can be developed to demonstrate environmental responsibility. The United Kingdom's CIA has latterly been promoting a combined environmental and health and safety management system that is developed under its Responsible Care initiative. More importantly though, both of the standards will, when fully implemented, provide an organization with a management-control system to constantly review and improve environmental performance. It is worth noting that ISO 14001 has a specific requirement for continual improvement in environmental performance whereas BS 5750 (ISO 9000) does not have such a specific reference with respect to quality. It looks like the ''wedge'' is built into ISO 14001. The continual improvement aspect of environmental management fits very well with the recent U.K. legislation on pollution control borne out of the Environmental Protection Act. This fit means that the development of a management system concurrent with regulatory driven improvement requirements will ensure that compliance with all the stated objectives can be demonstrated and, of course, will be auditable.

Robinson Brothers Limited was part of the pilot program for BS 7750 application within the chemical industry sector, now superceded by ISO 14001. Our industry-sector group feedback after the pilot stage and various changes that we discussed and proposed were actually incorporated into the 1994 revision of the standard and later into ISO 14001. Those changes, albeit only relatively minor, were the result of collectively about 30 years experience of ISO 9000 system operation. The voices of the standard users are being heard and can only fare well for future revisions and, in particular, to build on the experience of meshing the environmental management standard with regulatory requirements.

Health and Safety Management Systems BS 8800

The government department in the United Kingdom that looks after health and safety matters is the Health and Safety Executive (HSE). They have provided an excellent guidance document on health and safety management, namely HS(G) 65 "Successful Health and Safety Management." The HSE seems uneasy about the fact that the standards-writing authorities are seeking to develop a management-system standard. That unease was exemplified when at a meeting on integrated management systems, the HSE representative said that they could not support the achievement of third-party approval (certification) to the proposed standard. When asked to explain the response, there were words to the effect of "You should not control health and safety just to get a plaque on the wall." For an issue as important as workplace safety and welfare, the HSE is right. But who says you have to be externally accredited? Who will be qualified to audit you anyway; surely only the HSE for the present.

The result is that the standard, in practice, is referred to only as a Guide to Occupational Health and Safety Management Systems (i.e., there will still be no plaques to be gained). In the field of risk management and control, however, the guidance given by BS 8800 is quite comprehensive. The standard refers constantly to the aforementioned guidance document HS(G) 65 and draws parallels to both ISO 14001 and ISO 9000. In our case, we also refer continually to the guidance document HS(D) 65 and develop ISO 9000 style procedures to describe our health and safety management systems. We feel this is a means of ensuring that the all-important compliance check and improvement cycle takes place. All such procedures, as well as the EMS ones, shall be audited and reviewed to the same standard as the quality system. The corporate auditing team is charged with this responsibility.

Integrated Management Systems

So how do we run quality, environment, and health and safety management systems out on the shop floor? The answer is that we do already. However, it only meets one of the standard's requirements in full, namely ISO 9000. The most important place to achieve compliance is at the operator level. At this level, we can do the most harm to the product, the environment, and to people. The most important thing in our view is to control risk, which incidentally is the specific objective sought by the HSE. To that intention we have a program in hand to finalize our environmental and health and safety management system development that will culminate in the piloting of an integrated worksheet covering quality, environment, and health and safety for the chemical manufacturing process. I will not guarantee that it will be a single sheet of paper, but the key control requirements for the manufacturing process to demonstrate compliance with management-system requirements relating to material containment that will protect (1) the operator, (2) his colleagues, (3) the environment, and (4) the product quality will be afforded by this document.

Behind all this will be a set of procedures, mostly quite brief, and with flowcharting as appropriate to define specific elements of environmental or health and safety regulatory compliance. Sitting on top of the procedural hierarchy will be the management-system manuals for quality, environment, and health and safety describing in outline the respective management systems. You may think this does not sound very well integrated, particularly as separate quality, environment, and health and safety manuals will be required.

The present vogue for independent management-system certifications, modeled on similar lines, dictates that separate systems are adopted at the highest levels of the documentation hierarchy. This will be required at least until the recent initiative of the United Kingdom's CIA is widely adopted. Expertise in procedure writing among managers is still a major variable, but it is the operating staff who need clarity and simplicity in their day-to-day tasks if we want them to perform to the standards that we require. For this reason, integration starts at the bottom and permeates upwards.

Similarities between ISO 9000 and ISO 14001 do exist; namely appointed management representatives to control specific functions; work systems to be formalized; the need for an audit and review process; and management review meetings that should be held to discuss past, present, and future performance of the system. The operator knows what needs to be done to comply with quality, environment, and health and safety requirements. It is all stated on the operator's worksheet. All that remains is for the system to be managed effectively, and to be in line with whatever external standard requirements are necessary.

The experience gained with ISO 9000 means that organizations seeking to adopt other management-system standards will be also keen to adopt a similar language for operating within a standards framework. This, again, favors an integrated approach as senior management commitment can be facilitated by the use of a common language and a limited use of resources may be required in order to maintain compliance. Our experience with ISO 14001 is that senior management fully understand the need for formalized procedures to describe waste management and minimization activities. The need for the system to be auditable is well understood. So is the requirement for those managers to participate in a review of those audits and the general compliance of the organization with its declared environmental policy. This level of understanding and commitment could not have been achieved in the actual time experienced had we not gone through the ISO 9000 exercise.

The role of standards in industry has extended from the control of individual products to whole aspects of the business. ISO 9000 has set a standard for management-system development. There is little doubt in my mind that the ISO 9000 model will be around for some time and further management system spin-offs will probably emerge in the future. The main objective of standardized management systems is to provide a structured approach to managing the whole business or an individual business activity so that it may be seen to be effective by its compliance with a declared company policy. This, after all, is what organizations are developed for. It is a symbiotic relationship and the question to be answered is how can we hope to compete in a global market without a structure that will promote organizational efficiency and control operational improvement.

13

Strategic Standardization in Heavy Industry

This chapter is contributed by Keith B. Termaat, Cross Platform Closures Manager, and formerly Strategic Standardization Manager with the Ford Motor Company. The paper was first given at the American National Standards Institute (ANSI) Company Member Council Executive Committee (CMCEC) presentation as part of the World Standards Week Conference, Arlington, VA, October 1996 and subsequently published in the *Journal of the Society of Automotive Engineers* (SAE). ANSI, through the CMCEC and the Center for Strategic Standardization Management (CSSM) promotes the impact of standards and standardization on technology and innovation. After all, innovation is recognized globally as a powerful contributor to improving standards of living. But what is sometimes thought of as the flip side of innovation, standardization (the process) and standards (the performance), are also in the ascendancy as revitalized policy tools.

INTRODUCTION

To begin with, here are some items for thought on the standards and standardization process driven by trade, technology, and markets.

    1. Strategic standardization is not primarily about standards development. It is about deciding which standards ought to be developed and which ought to be used to satisfy objectives of your corporation.

    2. Standardization is not primarily a technical exercise. Standardization is a technical means to achieve policy/strategy objectives. This requires the exercise of diplomacy in achieving improved market share, profits, quality, and time to market.

3.   It follows that standards and standardization should not remain the sole purview of the technologists. The full benefits of standardization require that it enter the strategic planning process. It must be seen as an essential tool to achieving the aims of any enterprise.

The idea of standards is simple enough: a set of expectations of what we want to accomplish. Standards affect our lives every day. Transportation systems linking the global community, including motor vehicles, roads, bridges, tunnels, signage, safety, and air quality, are built on standards. Standardized "rules of the road" apply on land, at sea, and in the air.

FREE MARKET STANDARDS

But somewhere along the line we allowed the free-market standards process to detour into simply being a technical tool. Standardization often deteriorated into arcane debate on technicalities that soon drove out the very people who had the direction of the enterprise at heart. We are now returning to the fray, but many bring with us the old model of standardization. That is, the one about creating documents for buying commodities such as nuts and bolts and common parts. We need to educate the industrial world to the new standardization model, which must enter the planning process of the corporation and its individual business units, and we need to remind them that the model is global.

Let me share with you a few examples of how we are transforming Ford through standardization to make us leaner and more efficient. I will link that transition to strategic pressures on heavy industry with respect to international trade, technology, and markets.

THE FORD TRANSFORMATION

Eighty percent of the world's population live outside Ford's traditional markets. And our approach to these markets was fragmented. This made it more difficult to achieve our goal to be the world's leading automotive company. Ford believes that global standardization is one of the keys to reaching these potential customers. One of the most visible ways we are achieving our vision is by transforming Ford through a single set of worldwide processes and systems in product development, manufacturing, purchasing, and sales. Read standardization!

We used to view standardization as a technical exercise. Others saw it as a means to manage trade flows and to shape the direction of industrial development. Ford had to change and much of heavy industry must also change. Let's begin with challenging the common notion of standardization: the standardization of parts. There is nothing wrong with doing that; Ford is reducing horns from 33 to 3, batteries from 40 to 14, and so on; but that's Standardization 101. The new model is more strategic and proactive; and that model consists of a few very straightforward steps. These are steps you are already using in your businesses everyday, steps that parallel technological development as described below.

THE NEW STANDARDIZATION MODEL

1. Identify your policy and strategy areas of interest.
2. Formulate a position on each issue and then develop a standards proposition. That standards proposition is nothing more than a high-level solution, probably not more than a page or two, which outlines the essence of what you are trying to achieve. It describes specific technical issues that relate to your objectives.
3. Gain acceptance of that proposition inside your company and decide whether you want to go with external or internal standards. If internal, use the external steps internally.
4. Enter the fray in external standards organizations with a clear agenda in mind and with strategic direction every step of the way.
5. Influence selected standards bodies around the world and link up with allies whose interests align with yours. International Organization for Standardization (ISO), International Electrotechnical Commission (IEC), International Telecommunications Union (ITU), are global examples, but don't overlook Society of Automotive Engineers (SAE) and others you have worked with for many years. Their de facto global standards have become the benchmark around the world.
6. Selectively engage in the technical committees that will decide key underlying issues supporting your overall proposition.
7. Take part in the ratification and appeals process of the outcome.

This model avoids dissipating the energies of technical people on hundreds or even thousands of technical committees. The new model focuses instead on a small number of absolutely key committees. These committees are the senior policy fora in standards organizations around the world. They certainly would include industry associations—for us the United States Council for Automotive Research (USCAR) for technology and parts issues and American Automobile Manufacturers Association (AAMA) for policy issues. Ford is gradually working our way into senior positions on key panels to make sure that our interests are represented.

APPLICATION OF THE MODEL

The model is straightforward, but its application is not. Many of us would immediately leap to standardizing parts. We all do it, and we could probably benefit from doing it better. At Ford, we standardize fasteners, materials, and major parts, while improving customer satisfaction and product distinction. We are also working with GM and Chrysler to standardize plastics, fasteners and finishes, and Failure Mode and Effects Analysis (FMEA). There is an immediate payoff in commodity parts, and we understand that; standardizing parts has become an operating responsibility.

Notice that I slipped in an engineering method: FMEA. It is clearly not a part; but it is the subject of standardization. And it was strategically accomplished within Ford as a global process. We then obtained the agreement of GM and Chrysler to use the process in their practice. Subsequently, we all worked through SAE to have this become SAE J1739, which is now the standard for the "Big Three," and for the entire supply base.

Did standardizing FMEA across the Big Three save money? I don't know, because the costs are all indirect. But it did improve efficiency and quality, because analysis is so much easier and data can now be readily exchanged. Jaguar achieved better than 60% year-over-year quality improvement using FMEA and other standardized Ford methods.

The notion of standardizing practices introduces the notion of standardizing policies and strategies. It should not be too great a reach to see that other organizations have interests they seek to advance through standardization. Consider the ISO Committee on Consumer Policy known as ISO/COPOLCO. This policy committee has recommended standardization action on services, personal data protection, and energy efficiency. These are the same consumer standards experts who raised the first issues leading to ISO 14000, which, incidentally, Ford is implementing worldwide to standardize its operations.

But the interests of COPOLCO members do not necessarily align with those of industry. Consumers' priorities represent one voice among many. Our companies must develop positions on these important issues and participate in the process to be sure that our agenda is advanced. But what position do we develop? Let me give you some thought starters and examples with respect to international trade, technology, and markets.

TRADE

It should be no surprise that standards are being used to manage trade flows. Broad-based horizontal standards like ISO 9000 and safety standards now constitute potential trade barriers that have become the de facto price of entry into several countries. Any technical standard can become a barrier to trade. For example, the Japan Accreditation Board planned to enact a bundled software registration policy that would have required anyone selling products in Japan to reveal all the software secrets contained in their products. Led by Motorola, the United States defeated this attempt to erect a barrier to trade and compromise proprietary intellectual property.

But all standards actions are not as successful. The U.S. Fastener Quality Act, which was intended to ensure fastener quality from offshore suppliers, also applies to original equipment manufacturers (OEMs) and will cost the U.S. automotive industry an average of $18 per vehicle. These provisions, which rely on outdated quality-assurance measures and add no value are being contended by AAMA and U.S. fastener suppliers.

Trading in global markets very quickly exposes one to conformity assessment. That is, testing products to a particular standard and having to demonstrate conformity to some authority. It could be a regulatory authority, a customs authority, or a market requirement. The key notion here is to test once and get the test result accepted everywhere. That can only be achieved when testing and certification procedures are recognized around the world. Differences in testing procedures represent a significant impediment to designing vehicles and other products for global markets. Ford is active in negotiating a common recognition arrangement for conformity assessments around the world.

Make no mistake about it, technical standards are the new currency in managing trade flows around the world. Any standard, irrespective of its content, can be used to establish a trade barrier or remove one. Ford successfully concluded a negotiation with an Asian nation to accept our internal durability test in lieu of their government test. By removing this barrier, we entered that market six months earlier than otherwise possible.

TECHNOLOGY

Are there common threads of technology across a company's business and manufacturing units? Research and advanced engineering organizations often nurture core technical competencies in, for example, structures, mechanical devices, and electrical/electronics. These competencies identify areas of technology where it may be critical to enter the global standardization fray. If companies do not exploit core competencies, others will write the rules by which they will have to play. Being technologically preeminent, while critically important, is no assurance of success. Apple Computer is the developer of the user-friendly MacIntosh operating system. Not licensing this standard to others has, according to *Business Week*, cost them $20 to $40 billion in revenue.

Ford's creative vehicle configurations are the basis for our competitive success. But different may not always be better, since we also reduced our vehicle platforms and our powertrain combinations by one-third. So the first step of strategic standardization is to reduce the number of variations in products without limiting customer choice. Mabuchi Motors, a maker of small electric motors for transportation and other sectors, serves 70% of its customers with only 20 products.

Another challenge is to raise standards for marketplace advantage. Ford raised its internal technology performance standards for durability, corrosion resistance, and climatic extremes. Then we applied these new standards to all our vehicles, worldwide.

Standardized technological solutions make life easier. Life for mechanics in the service bay was improved by a standardized Service Bay Diagnostic System that is an industry leader. Ford also came up with a fastener that cannot be cross-threaded when it's driven with power tools. This invention led to a standardized solution in assembly operations. Major improvement in cost and quality resulted.

Ford and other OEMs are actively developing electric vehicles. Whole new families of standards for new technologies must be established. The automotive sector is involved; and for the first time, the electrical utilities sector. The stakes are enormous. Should the current home electrical infrastructure standard be preserved? If so, which nation's? Should the vehicle be compatible with it? Or should we go to a new method for charging vehicles requiring an expensive installation in every home? Different players have different views. (The French favor solutions to benefit their standard infrastructure and their industries.)

What is the point? To remain competitive, we must play across many industry sectors simultaneously and we must play strongly in our core technologies so that others do not write the rules by which we play.

MARKETS

Ford Automotive Operations consists of single, centralized departments and processes; one for sales and marketing, one for manufacturing, and one for purchasing. Standardization is managed as a global responsibility, as are all other functions. Nontraditional markets require new and tougher product standards and require knowledge of regulatory standards. So we have teams operating around the world to identify what is required, not only in road infrastructure and fuel quality, but also in driving habits and local standards. This research data has led to new internal product-acceptance standards.

This makes an important point. Every standardization action requires a decision whether to go with an internal standard or an external standard. At Ford, we are moving to external standards except where we have a proprietary interest to protect. Then we maintain the standard internally. Product acceptance standards are proprietary; they are what make a Ford a Ford.

Global markets require reliable global partners; and the partners change with the issue. We participate with General Motors and Chrysler, for example, in the Partnership for a New Generation Vehicle with the opportunity to work on standards. This cooperative venture between the Big Three and U.S. government research laboratories seeks technological breakthroughs to assure the U.S. auto industry remains fully competitive. This initiative is a partnership with the U.S. government, but as a globally integrated company we must also pay attention to the interests of other governments.

At Ford, we have a finely honed process for identifying what's happening in the regulatory standards world and this process is now becoming global. We connect into leading capitals to tell us what is going on. We are now painstakingly building the same intelligence capability with respect to nonregulatory standards. Increasingly, regulations may be based on voluntary standards.

The United States passed the National Technology Transfer and Advancement Act of 1995 (P.L. 104–113), which directs regulatory agencies to rely on voluntary standards as the basis for regulations. So, your interests written into voluntary standards provide an important leg up on a potential regulatory result.

NEXT STEPS

But what standardization actions are potential next steps for heavy industry? First, do not create other standards-developing organizations; the world has too many as it is. Do select strategic issues.

1. Select preregulatory issues for consensus and current regulatory issues for harmonization in the standardization infrastructure.
2. Standardize material composition and gauges for aluminum, steel, and nonmetallic materials.
3. Standardize test methods. Support a global process for accrediting and recognizing laboratories to ISO Guide 25.
4. Use standards and standardization as tools to open markets by participating in ANSI, ISO, IEC, and, of course, SAE.
5. Identify a handful of advanced and core technologies that you can unite around; then standardize. With respect to the last item, let me describe the next generation motor vehicle and its impact on standardization.

PARTNERSHIP FOR A NEW GENERATION OF VEHICLES

The government and private Partnership for a New Generation of Vehicles (PNGV) illustrates the importance of standards and standardization to new technologies that require whole families of new standards. Listed in the following text are some representative short- and long-term technologies that the PNGV must have for success. The list is not all inclusive, but has a tremendous impact on the success of PNGV. The first five prioritized technology areas, particularly significant to the PNGV program, are classified as high overall potential regardless of schedule (long-term potential beyond 2004).

1. Lightweight structural materials such as aluminum, composite, and steel.
2. Advanced manufacturing such as simulation and modeling, systems/software integration, agile/flexible manufacturing, alternative processes, test methodologies, and current process optimization.
3. Energy Conversion such as four-stroke, direct-injection engines, fuel cells, and turbines.
4. Energy storage devices such as battery/ultracapacitor, power electronics, and battery capture of lost mechanical energy to offset transient power demands. Longer-term ideas include ultracapacitors and flywheel.
5. Efficient electrical systems to improve vehicle performance through fast control responses, and fuel economy through battery charging, electric steering assist, active suspension, variable-speed AC compressors, electric propulsion, and regenerative braking.

LONG-LEAD TECHNOLOGIES

Long-lead technologies include the following:

1. Waste heat recovery;
2. Reduction of mechanical losses;
3. Aerodynamics/rolling resistance improvements;
4. Improved efficiency of internal combustion engines;
5. Emissions control;
6. Fuel preparation, delivery, and storage; and
7. Interior thermal management.

The next major milestone is the selection of the most-promising technologies by 1997. The intent of PNGV is to develop production prototype vehicles by 2004.

SUMMARY

I have given you some thought starters on the standards and standardization process driven by trade, technology, and markets. Remember the following salient points.

1. Strategic standardization is not primarily about standards development. It is about deciding which standards ought to be developed and which ought to be used to satisfy objectives of your corporation.

2. Standardization is not primarily a technical exercise. Standardization uses technical means to achieve policy/strategy objectives. This requires the exercise of diplomacy in achieving improved market share, profits, quality, and time to market.

3. It follows that standards and standardization should not remain the sole purview of the technologists. The full benefits of standardization require that it enter the strategic planning process. It must be seen as an essential tool to achieving the aims of any enterprise.

CONCLUSION

At Ford, standards development has become a competitive venture—fast paced, intellectually demanding, results oriented, and global. It is all about the historic notion of setting high standards. At Ford, setting high standards is an essential principle; high standards for ourselves, for our company, and for our customers. Please join us to develop high global standards to our mutual advantage, which will advance our industry in the second century of automotive transportation.

REFERENCE

1. National Research Council. Review of the Research Program of the Partnership for a New Generation of Vehicles. Washington, DC: National Academy Press, 1996.

14

The U.S. Standardization System: A New Perspective

This invited chapter was prepared by Robert B. Toth, President of R. B. Toth Associates, Alexandria, VA. Bob Toth has earned an international reputation and is a recognized standardization management expert. R. B. Toth Associates specializes in strategic services for the industrial infrastructure, and provides consulting services in all aspects of standardization and conformity assessment to clients worldwide: international agencies, manufacturers, standards developers, trade organizations, and government agencies. He has assisted more than a dozen countries to restructure or upgrade their metrology, standardization, and conformity assessment systems. With permission of the author, the chapter was specially adapted from a prior publication for inclusion in this book. Bob Toth is the author and editor of many papers and books including *Standards Management: A Handbook for Profits; The Economics of Standardization*; and *The Role of Standardization in Economic Development*. He is a graduate engineer, certified standards engineer, Fellow of the Standards Engineering Society, and Member of the American Society of Mechanical Engineers. He can be reached at bob.toth@erols.com

BACKGROUND

The odds are very high that an American attending an international standardization meeting or consulting in a foreign country will be asked about the U.S. standardization system. How is it organized? Who is responsible for developing standards? How many standards-developing organizations are there? How many standards? Who sees to their implementation? What is the government's role?

131

Why is there more than one standard for many commodities? Foreign engineers who are used to dealing with their national standards institute can be very critical of the decentralized U.S. system. Many are dissuaded from applying U.S. standards because of real or anticipated problems in choosing, obtaining, and applying them. Many Americans raise similar questions. Some think that all standards come from the government. While most U.S. standardization specialists are familiar with the standards in their fields, few have an overview of the U.S. standardization system.

The very complexity of the U.S. standardization system is perceived by many global trading partners as an inherent trade barrier. When U.S. trade negotiators press for openness and transparency the rejoinder is that the U.S. standards system is more restricted and less accessible because it is so convoluted (*"le labyrinthe des normes!"*). Some are not only complaining, but recommending that the United State, restructure its standardization system:

> . . . it is absolutely necessary to demand in-depth reforms of the American system to make it more transparent, more consistent and more in conformity with international standards. Why not insist that the Americans consider the introduction of a system similar to the [European Union's] New Approach? That system is flexible and gives an equal share of responsibilities to the private and the public sectors. Many American companies would welcome such a development (1).

To alleviate these complaints and starting in 1984, the National Institute of Standards and Technology (NIST) has published *Standards Activities of Organizations in the United States*. This is a directory of U.S. national mandatory and voluntary standardization activities that specifically addresses the concerns of those who want to know details about U.S. standards development. The directory has become one of NIST's most popular publications and is widely distributed in the United States and abroad. Information accumulated for the more than 700 entries in the 1996 edition (2) is the primary source for this overview, together with material R. B. Toth Associates has gathered for a new NIST publication, *Profiles of National Standards-Related Activities*. This describes the metrology, standards, testing, quality, and accreditation operations of almost 100 countries. *Profiles* can be accessed at http://ts.nist.gov/ts/htdocs/210/217.htm

SOME BASIC DATA

The U.S. standardization system is continually changing and any profile is necessarily a snapshot of a moving target. At the national level the United States currently maintains about 93,000 standards in an active status. Figure 14.1 shows

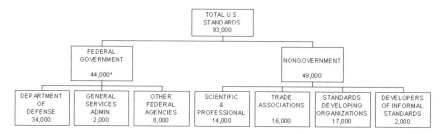

Figure 14.1   Estimated number of U.S. standards and their developers in 1996. In addition the federal government has adopted 9,500 nongovernment standards.

the various categories of standards developers and their output in numbers of standards, as of 1996. In addition, the federal government has adopted 9,500 nongovernment standards.

In the past five-year period the total number of U.S. standards has not increased. And, for the first time, the private sector maintains more national standards than the government. Federal agencies are canceling more standards than they develop, and are also adopting nongovernment standards instead of preparing their own. The total number of federal government standards has decreased by 8,400 in the past five years, and the number of adopted nongovernment standards has increased by 4,600. Nongovernment standards-developing organizations are adding to the overall body of standards at a net rate averaging 3.4% per year. Very few voluntary standards are being canceled or inactivated. Significantly, this increased output can be attributed primarily to midsize standards developers. In 1996, the standards of the 20 major nongovernment standards-developers shown in Table 14.1, constitute 71% of the nongovernment standards database. In 1984, their standards made up 81%.

The number of published standards is not necessarily an absolute indicator of overall activity level or significance. Although counted as a single standard, the 12,000-page Boiler Code of the American Society of Mechanical Engineers, and any of the model building codes are impressive in size and importance. Thus they're not comparable to 95% of the other standards published in the United States.

THE STANDARDS DEVELOPERS

The number of federal agencies developing standards has remained relatively constant. The number of private-sector standards-developing organizations with

Table 14.1   Twenty Major U.S. Nongovernmental Standards Developers

Aerospace Industries Association	3,000
American Association of Blood Banks	500
American Association of State Highway & Transportation Officials	1,100
American Conference of Govt. Industrial Hygienists	750
American National Standards Institute (as ANSs)	1,500
American Oil Chemists Society	410
American Petroleum Institute	500
American Railway Engineers Association	400
American Society for Testing and Materials	9,900
American Society of Mechanical Engineers	600
Association of American Railroads	1,400
AOAC International	2,100
Cosmetic, Toiletry and Fragrance Association	800
Electronic Industries Association	1,300
Institute of Electrical and Electronics Engineers	680
National Association of Photographic Manufacturers	475
Semiconductor Equipment and Materials International	450
Society of Automotive Engineers International	4,550
Underwriters Laboratories	780
U.S. Pharmacopeial Convention	5,000

active, ongoing standards-development programs has increased by about 50 organizations developing formal standards, and a whole new category of standards developers has emerged. These are organizations that, for the most part, work outside the traditional standards-development framework. Figure 14.2 depicts the numbers and types of organizations that develop national standards. These can be categorized into four groups. The first three prepare formal or de jure standards.

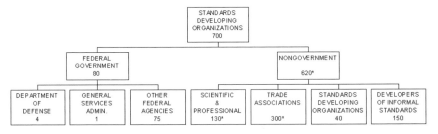

Figure 14.2   Number of U.S. standards developing organizations. Approximately 130 of those marked with * have prepared a few standards in the past and occasionally update them, but are not actively engaged in ongoing standards development.

1. *Scientific and professional societies*: Some of the more than 2,000 societies in this group that, in addition to their professional and educational roles, also develop standards. Examples are the Association for Computing Machinery, the Institute of Electrical and Electronics Engineers, and the Acoustical Society of America.

2. *Trade associations*: These associations deal with mutual business problems in a particular industry, and promote the industry and its products. To address their objectives, many trade associations develop standards for the products manufactured by their members, although a few concentrate on developing standards for products used by their industries rather than the products they supply. The Aerospace Industries Association and the American Petroleum Institute are two trade associations that develop standards for items used by their member companies.

3. *Standards developing organizations*: Organizations founded specifically to develop standards are often designated *standards developing organizations* (SDOs). Some confusion occurs when the four types of organizations are collectively called SDOs. The oldest SDO in the United States is the U.S. Pharmacopeial Convention. It published standards for 219 drugs in 1820. The American Society for Testing and Materials can be classified as an SDO, as can many of the organizations that serve as secretariats for ANSI-accredited committees (e.g., the Alliance for Telecommunications Solutions).

4. *Developers of informal standards*: This category could be considered a subset of the group of SDOs founded specifically to develop standards, but these SDOs operate outside the traditional standards-development framework. They develop or promote standards that are described as ad hoc, de facto, or consortia standards. Another apt descriptor is that they develop ''limited consensus standards.'' It is estimated that there may be as many as 150 such organizations in the United States and 50 to 70 others in the rest of the world.

INFORMAL STANDARDS DEVELOPERS

Until recently, it was fairly easy to identify private-sector standards developers. Their formal, de jure standards are developed using processes that are open and that apply the principles of due process to achieve decisions through consensus. Procedures for assuring openness enable various interests not only to identify organizations that might be developing standards that could affect them, but to communicate with the standards developers and to receive notices of meetings, agendas, and draft standards. In recent years, however, many new nongovernment standards developers have emerged who do not follow traditional standards-developing procedures. For whatever reasons, often expediency, these standards developers ignore or short cut the traditional procedures. The lack of openness makes it very difficult to identify the organizations that are developing standards in this manner, and, even when an ad hoc group or consortium is identified, it is difficult to communicate with an organization that has no formal structure or fixed secretariat.

A standard is not an end in itself. It is an intermediary and a medium for communications. Organizations want to buy products and services, not standards. User demands drive markets and the markets should drive the standards, but standards have not kept pace with the markets in telecommunications, information technology, and other rapidly developing fields. Technologies with market lives measured in months cannot wait years for formal standards to emerge from traditional processes. The traditional developers recognize the need to provide more timely standards, but must balance that need with their obligation to maintain the right to due process by all materially affected interests. Consortia and ad hoc groups are more focused on achieving near-term results, and their standards are developed without heavy emphasis on formal due-process procedures. In addition, the members of consortia and ad hoc groups are relatively homogeneous and single minded in their goals. Characteristics of formal and informal standards developers are shown in Figure 14.3. Informal standards can be organized within two general categories, proprietary and consortia, as shown in Figure 14.4.

Proprietary Standards Development

When the products or services of one company become widely accepted as ''the standard'' within a market, the result is a proprietary informal standard. There are two ways that such standards emerge:

1. Proprietary de facto standards: A company strategically positions its products to expand market share and collect royalties by licensing intellectual property rights. Examples: Adobe Postscript, Netscape Internet Browser, and Microsoft Windows.

Type	FORMAL STANDARDS	INFORMAL STANDARDS
Type	*de jure*, consensus, industry voluntary	*de facto*, *ad hoc*, consortia, user groups
Participants	diversified, varied objectives	relatively homogeneous, single minded
Procedures	due process, open, consensus, public review, authorization	ignore or short cut traditional procedures; expediency; near-term results
Communication	easy; fixed address; standards readily available	very difficult; often no formal structure, secretariat, or fixed address; often restricted distribution

Figure 14.3   Characteristics of formal and informal standards development.

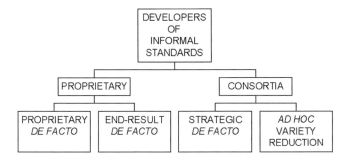

Figure 14.4 Categories of formal and informal standards.

2. End-result de facto standards: Market forces designate one company's product as the standard from among equally effective competitors. Example: NTSC VHS-format video recording (versus Beta).

The standardization community has little direct linkage to proprietary standards development that is centered on one company and its product. Proprietary informal standards and their developers are not included in the totals presented in Figures 14.1 and 14.2.

Consortia Standards Development

This classification includes standards developed by ad hoc groups of suppliers or users, research and development consortia, and patent licensees. These groups work together to develop informal standards or to select those standards that will be given preference within a particular sector or market. Motivations of the sponsors differentiate the two classes of consortia standards described in the following text.

3. Strategic de facto standards: Suppliers, and occasionally users, attempt to establish sufficient critical mass to define the standard or standards in a particular field. Examples include Dolby, Notes Consortium, Advanced Photographic System, and DVD Multimedia Discs. Strategic De Facto Consortia standards represent probably more than 95% of the informal standards in use today. Most of these are supplier driven and define software and connectivity requirements for information technology and telecommunications.

4. Ad hoc variety-reduction standards: Traditionally, the application of industrial standardization principles has focused on reducing the number of types, sizes, and kinds of parts, materials, and processes to realize cost savings and improve productivity. In some sectors, users complain of an overabundance of standards because standards developers are not sufficiently selective, but publish

standards for nearly every product offered in the market. To fully realize the potential of standardization through economies of scale, industry consortia have formed to select and harmonize existing standards for preferred use within their industries. The United States Council for Automotive Research (USCAR) is a major consortium with a Strategic Standardization Board working in this way. Other user consortia recognize that many common practices exist within their sectors that could benefit from industry-wide standardization. Most of these practices are documented as internal company standards. Rather than processing these internal standards through the traditional, formal standards-development process, they prefer to prepare harmonized standards that are intermediate between company standards and formal consensus standards. The Process Industry Practices (PIP) initiative exemplifies this approach (see Chap. 19).

Within Europe, and some regional and international SDOs, informal standards can be designated Publicly Available Specifications (PAS) if their developers make them available for public comment and meet a few other criteria. These PASs can then be fast tracked to emerge as formal standards. The X/Open Consortium and European Workshop on Open Systems (EWOS), for example, have been designated as Recognized Submitters of PASs by International Organization for Standardization/International Electrotechnical Commission (ISO/IEC) Joint Technical Committee 1, which enables the committee to adopt or "transpose" these informal standards.

EVOLUTION

The developers of informal standards start out with simplified fast-track procedures, but some find that the traditional due-process procedures are necessary to accommodate the needs of their members. Others, upon developing a standard, have no mechanism for maintaining it. Some consortia recognize that acceptance of their proposals could be more widespread if their informal standards received a technical review by experts and assurance that the consortium's standards would be integrated into the existing body of formal standards. For these and other reasons, an increasing number of consortia are acting more like traditional standards developers, while others are affiliating with developers of formal standards. VMEbus and the Consortium for Advanced Manufacturing International are ANSI accredited. The Document Management Alliance is now a task force within the Association for Information and Image Management. USCAR has a working relationship with the Society of Automotive Engineers (SAE). And as consortia mature, many evolve into full-service trade associations—the same path that some well-known standards developers took in years past.

Initiatives to promote the acceptance of proprietary and consortia-developed informal standards are increasing general awareness of the role of standards. While traditional standards developers complain that groups developing informal standards are interlopers, many traditional developers are reengineering their procedures to be more responsive to special interests. There are no indications, however, that the activities of nontraditional standards developers will decrease. Marketing experts have adopted standards development as one more essential stratagem they can use to attain their objectives.

GOVERNMENT STANDARDS

As Figure 14.1 indicates, the Department of Defense (DoD) is by far the largest developer of standards in the United States. About 8,000 of its 34,000 standards define strictly military commodities and practices. The remainder define parts, materials, industrial equipment, and consumables, many of which are used by commercial industry. The 5,650 Federal Specifications and Standards and Commercial Item Descriptions prepared by DoD and the General Services Administration (GSA) are widely used by federal and state agencies to procure many common products, and federal test methods are used in many industries. It is estimated that there are another 8,000 standards developed by such federal agencies as the Occupational Safety and Health Administration (OSHA), the Environmental Protection Agency (EPA), the Food and Drug Administration (FDA), National Institute of Standards and Technology (NIST), and others. Many of these are in a gray (or overlapping) area where it is difficult to differentiate between standards and regulations.

Recent trends in DoD standards development have been toward the elimination of detailed military specifications and standards in favor of adopting nongovernment standards or developing commercial item descriptions and defense performance specifications. From 1990 to 1995, DoD canceled over 6,700 military specifications and standards, while adopting 2,400 nongovernment standards and issuing nearly 1,900 commercial item descriptions for procurement of off-the-shelf products. The DoD has now adopted over 7,400 nongovernment standards and published over 4,200 commercial item descriptions. Other agencies are also using nongovernment standards extensively. GSA has adopted more than 800 nongovernment standards. The Department of Energy's (DOE) Technical Standards Program applies 60 DOE standards and 800 from the private sector. It is probable that a number of federal agencies have adopted the same nongovernment standard, so the total number adopted by all federal agencies may be overstated.

TRENDS AND CHANGES

The increasing role of organizations that develop standards outside the traditional framework is not the only trend that has affected the overall number of standards and their application. Harmonization of standards used in Canada and the United

States is resulting in consolidation and simplification. The outcome of these efforts is one or two standards where previously there were three or four Canadian and half a dozen U.S. standards. The national adoption of international standards is increasing. Ten years ago, fewer than ten national adoptions (indicated by the designation ANSI/ISO) were listed in the ANSI catalog. Today the number of adopted ISO and IEC standards exceeds 200. Many of these are based on U.S. standards that can be canceled. Others preclude the need to develop U.S. standards. National adoptions, however, are not a true indication of the extent that U.S. manufacturers and service providers are implementing international standards. Unlike many other countries, U.S. standards developers and users, particularly in high-technology industries and global markets, do not wait for the ANSI imprimatur before citing an international standard. It is estimated that more than half of ISO's and IEC's standards are implemented directly by U.S. standards developers and users.

The U.S. standardization community is in a period of change and, within some industrial sectors, redirection. Government contractors, particularly within the defense industry, are learning to cope as the DoD implements policies that promote the use of performance specifications and nongovernment standards, canceling thousands of military standards and specifications. Private-sector standards developers are stepping in to fill apparent voids. Equipment suppliers and users are working in many forums to develop standards to promote interconnectivity for telecommunications and information technology systems. A number of trade associations have instituted standardization initiatives to accommodate their members' needs for electronic data interchange (EDI) capabilities.

At this time, many more organizations prefer to describe their publications as guidelines or recommended practices rather than as standards and a new class of standards is being published for general use: Interim or Draft Standards. Many organizations are responding to the needs of new technologies by developing standards that can be implemented on a trial basis prior to publication of a fully definitive, final standard.

Many organizations have added *international* to their name. A dozen use only an acronym, so that their fields of specialization are not evident to the uninitiated. Examples include AIM USA, AOAC International, NACE International, NSF International, and SFI Foundation. A more positive trend is the increasing availability of e-mail addresses that can be used to communicate with standards developers in government and the private sector. Each year more organizations announce that their World Wide Web (WWW) sites are on the Internet providing access to information about their standardization programs. Many make it easier to order standards by putting their catalogs online. The Defense Information Systems Agency and the DoD Single Stock Point enable some users to download standards from the Internet. A few standards committees are developing standards online, but the most extensive user of the Internet for standards development and delivery is, appropriately, the Internet Engineering Task Force. Thousands of experts develop consensus positions on a range of technical issues in the transparent, open forum of the Internet.

REDUNDANCY, OBSOLESCENCE, AND ECONOMICS

One characteristic of free enterprise is free entry into the marketplace. Another is once a market has been created, the entrant wants to retain as much of a share as possible by discouraging new entrants. These two characteristics combined with the American proclivity for establishing and joining organizations are the basic reason for overlapping and redundant standards. For example, fourteen organizations develop and publish standards in the field of ventilation and air control. Anyone involved with air flow and control has the problem of trying to decide which of the many similar standards published by these organizations should be used. Often the customer must decide which standards to invoke in a purchase contract. Faced with this dilemma, the customer can be very critical of our standardization system.

A relatively small portion of the 35,000 standards prepared by the twenty major standards developers are redundant or overlapping. The portion is much higher, however, for the remainder. The most apparent sector is building and construction. One reason is that standards and specifications are the pervasive media for defining procurement requirements. More than 11,000 standards (12%) are applicable to building and construction. More than half of the nongovernment developers of formal standards prepare standards used in building and construction. Two hundred of the 300 trade associations that prepare standards do so exclusively for building products. Another factor is the multiplicity of building codes. In this country there are a minimum of 14,000 building codes. Some estimates are as high as 20,000. While the three model building codes have brought some order and uniformity within their areas, most state and local jurisdictions merely use these as the starting point on which to tack amendments and additions.

The number of available standards may not be indicative of their value to industry and commerce. Real value is realized only when standards are used. Methods have yet to be developed to measure how often, or how widely, individual standards are implemented. Standards developers tell us that 80% of their orders for individual standards are for 15 or 20% of the total number of published standards. Many standards are seldom used. We also know that, in spite of five-year reviews, a substantial portion of our standards refer to and document obsolescent technology that, while appropriate for spares and maintenance of older equipment, is no longer appropriate for new designs. Some believe that 25 to 30% of our national standards—both government and industry—fall into this category. Standards developers are reluctant to withdraw outdated standards or designate them "Inactive for New Design." The net result is that users complain that there are too many publications called *standards* that are not deserving of the title. This proliferation dilutes the impact and effectiveness of those documents that truly deserve to be called standards.

There is urgent need to focus on the "bottom line," in other words, on net profit and return on investment, to be more critical in initiating new standardization projects and to manage resources better so that necessary standards are available when needed. The standards community has little information on the degree to which standards are being applied and no information on the amount of resources that go into standards development. Some basic research is needed to determine the extent to which standards are being applied. Criteria should be developed to assist in determining when there are sufficient standards in a particular field. Too often resources are invested in standards projects that have limited or questionable payoff. The benefits of standardization are derived by the users of standards, not by standards-developing organizations. Standards developers and their committees should reach out beyond the committee members to determine the real needs of standards users. The proven techniques used in market research are particularly useful for this purpose.

A GLOBAL PERSPECTIVE

On a global basis, the most noteworthy development is the shift in focus in many countries for policy, direction, and coordination of standardization and conformity assessment from the national standards body to the national accrediting authority. The World Trade Organization (WTO) agreement on Technical Barriers to Trade (TBT) has provided the impetus to create an efficient, transparent, fair, and harmonized means of international acceptance of traded goods. To support domestic industries and promote trade, countries have established, or are in the process of establishing, a national infrastructure for accrediting, testing, and calibration laboratories, inspection bodies, standards developers, and certifiers of agricultural and industrial products, personnel, quality systems, and other standards-related activities. Countries are implementing the concept of "one-step conformity assessment" not only to minimize costs, but to achieve recognition by other nations. The national accrediting authority fills the need for a recognized body empowered to formalize agreements with regional organizations and other nations. This development, which focuses on implementation of standards rather than their preparation, coupled with the enhanced role of regional and international standards organization, has prompted some national standards bodies to reassess their role and organizational structure.

The nations of the world currently maintain approximately 800,000 national standards, an estimate based on inputs for the NIST directories and approximations for countries that did not provide data. A large portion of these standards have common technical requirements and differ primarily only in language. In addition, the standards of most countries reflect the requirements of international standards. National standards bodies report that an average of 43% of their standards are identical to, technically equivalent, or based on international standards.

The United States has a plethora of standards. Figure 14.5 lists some of the major collections of standards in the world (3). To achieve an apples-to-apples comparison, DoD's 8,000 standards for military products mentioned previously should be subtracted from the U.S. total for an adjusted total of approximately 85,000 standards.

Most countries have centralized standardization systems within a single standards body. However, it may be a surprise to some that the standardization systems in many countries are not monolithic. Some of the larger countries have numerous standards-developing organizations in addition to the primary standards body. In Germany for example, 153 private-sector standards developers have prepared 15,000 standards, while DIN has prepared 22,000. In Japan more than 200 trade associations and professional societies prepare standards within

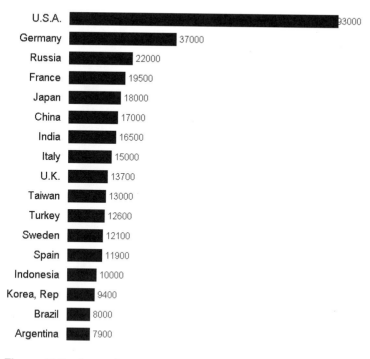

Figure 14.5   Some of the world's major collections of standards.

the JISC framework. In France, thirty semiautonomous, sectorial bodies (Bureaux de Normalisation) prepare industry-specific standards while AFNOR prepares cross-cutting horizontal standards.

In these and other countries, trade associations and technical societies are an integral part of the national standardization system. In the federated U.S. system, there is considerably less coordination among the standards developers and more competition for resources and influence. Because U.S. standards developers are more dependent on the sale of their publications to support standards activities than foreign counterparts, there is much more emphasis on the sale of the standards, rather than expanding markets for the products defined by the standards. A few trade associations are reexamining their cost-recovery policies as they realize these policies may not support the long-range objectives of their member companies.

Federal agencies coordinate their standardization activities through the Interagency Committee on Standards Policy that was rechartered in 1994 to implement revised federal policies for using voluntary standards specified in Office of Management and Budget (OMB) Circular A-119. Coordination of the nongovernment standardization system has been the responsibility of ANSI for nearly 80

Table 14.2   Contrasting Approaches to the Development and Application of Standards

USA	Other OECD Nations
Distributed	Centralized
Pragmatic	Systematic
Reactionary	Anticipatory
Inch-Pound	Metric
Tolerated; implementation questioned	Acceptance; immediate implementation
Maximize role of private sector	Responsive to government direction and national policy
Focus on market needs and strategic uses	Aim to influence regional and international standards
Entrepreneurial and individualistic	Tools of industrial policy
Open and transparent	Often closed, negotiated standards development
Appeals mechanisms exist	Appeals procedures are exception
Use whatever standard is appropriate	Precedence to designated national standards
Direct application of international standards	Adoption or transposition of international standards
Minimal regional standards development	Considerable investment in regional standardization
Self-certification and warranties	Third-party testing and certification

years. ANSI's primary standards responsibilities are to accredit standards developers, approve standards for designation as American National Standards (ANSs), and represent the United States in ISO and IEC.

More than 180 standards developers are accredited by ANSI and their standards constitute more than 75% (37,000 estimated) of the country's nongovernment standards. Yet many of these accredited standards developers process only a few of their standards through ANSI for the ANS designation. Currently, there are just over 11,000 ANSs. Standards users in the United States and abroad are understandably confused: What is a *national* standard? Does an ANS have precedence over other standards? Does the ANS designation afford any special legal or market advantage? This is just one area that raises questions about the U.S. standardization system, and prompts comparisons with the systems of other countries. However, the U.S. system is unlike any other in the world because there are basic cultural and economic differences that have determined how we approach standardization. Table 14.2 summarizes some of the more salient differences.

PRIME MOVERS

In the United states, a growing number of major corporations have become aware of the strategic nature of standards and standardization. Many have encountered trade barriers in the guise of standards. This recognition has prompted a significant number—probably 20 to 30% of the Fortune 1000—to reexamine their internal standardization programs and their influence on those external standardization activities that impact their businesses. As a result, there is more high-level attention on domestic and international standards developers. Pragmatic corporate managers consider standards developers as just another group of service suppliers. They allocate their resources to those SDOs that provide the best value for the investment. The largest bloc of standards users, in other words, industry, is pressuring standards developers to be more responsive. They point out that industry not only provides the volunteers to develop standards, but industry is the major purchaser of those standards. In recent years, industry has reengineered, downsized, outsourced, and focused on customer satisfaction so that it could be more productive and competitive. These standards users expect similar initiatives from standards developers.

Corporate America's critical examination of standardization is not limited to domestic standards developers. All of the larger corporations have worldwide markets and most are shifting from being international corporations to global enterprises. International corporations have been functioning basically as numerous individual companies operating under one name. The global enterprise is borderless, with free movement of peoples, ideas, and products, so that resources can be used quickly and effectively to bring products to market. When global enterprises examine the process by which international standards are developed, they conclude that the system is anachronistic and inherently ineffective. They are particularly critical of the principle that standards development and approval is accomplished through national delegations and the votes of national standards bodies. In their view, most standardization is accomplished on a sectorial basis. When global enterprises need global standards they believe that the best course is to work together directly and not through surrogates or intermediaries. A few corporations in well-defined industrial sectors characterized by a limited number of large, competitive global enterprises are starting to float this concept among key decision makers.

Another reaction to the inefficiencies of the standards development process at the international level is to withdraw from the process. Supporters of this approach are rallying round the following manifesto:

> Abandon the commitment to international coordination. The vast majority of high quality standards are being set and refined in the US. The notion that we should coordinate in an international effort is misguided and a drag on US efforts. We are the largest market, and while we should be sensitive to opening foreign markets, we should do so with our standards and stop wasting efforts to educate the world and incorporate what in the last analysis are mostly vain national efforts to impact our standards (4).

This isolationist approach, some would say jingoistic, while radical is indicative of real concerns about real problems. It also reflects a high level of frustration and skepticism that substantive improvements can be accomplished within the existing system.

WHAT'S NEXT

There is little likelihood that the U.S. standardization system will adopt the European model. It is evident, however, that the dynamics of standards development (and to an even greater extent conformity assessment) are changing. In the past five years, the rate of change has quickened. The next five years will certainly see more changes.

We will probably see the emergence of a few developers of sectorial global standards outside of the ISO and IEC framework. Then again the ISO and IEC are starting to provide opportunities for affinity groups of global enterprises to work directly, rather than through national bodies, to develop their global standards.

The dichotomy between the distributed U.S. standardization system and the hierarchical, centralized systems of most other countries will probably widen as other countries harmonize more of their national standards as prescribed by central coordinators within regional blocs. The flexibility of the U.S. system and its ability to assimilate a wide range of standardization processes will mirror the dynamics of a global economy. In contrast to the rigid, authoritarian structure that characterizes the standardization systems in other parts of the world, an amorphous, but responsive, system might better serve the needs of global enterprises and other users, and demonstrate the competitive advantage of America's approach to standardization.

Acknowledgments

Special thanks are due to the National Institute of Standards and Technology, Office of Standards Services, and particularly Maureen A. Breitenberg who managed the contract that enabled the author to acquire much of the data used in this paper. However, interpretation of the data, and related observations, conclusions, and opinions are solely the responsibility of the author and in no way reflect the views or policies of NIST.

REFERENCES

1. F Nicolas. Deciphering. In: Enjeux. Paris: Association Francaise de Normalisation, June 1996.
2. RB Toth, ed. Standards Activities of Organizations in the United States. National Institute of Standards and Technology, Special Publication 806, 1996 ed. CODEN: NSPUE2. Washington: U.S. Government Printing Office, September 1996.
3. Newly independent countries formerly associated with the Soviet Union generally have large numbers of standards, for example, Ukraine 21,000; Belarus 19,000; Poland 15,400; and Bulgaria 13,000.
4. S Oksala, A Rutkowski, M Spring, O'Donnell. The Structure of IT Standardization. StandardView: March 1996.
5. B Collins and W Leight. Setting the Standards. Mechanical Engineering, vol. 122, no. 2, 46–53, February 2000.

15

The Modern-Day Archimedes: Using International Standards to Leverage World Markets

This paper by Stephen C. Lowell, Standardization Program Division, Office of the Secretary of Defense with the U.S. Department of Defense (DoD) received first place in the 1997 World Standards Day Paper Competition, in which the overall theme that year was "Standards: Builders or Barriers to Trade." It was published subsequently by the Standards Engineering Society (SES) in its journal *Standards Engineering* Vol. 49, No. 6, November/December 1997. Stephen Lowell is Director of the SES Technical Council and a Fellow of the Society.

The Greek mathematician and scientist Archimedes once boasted that given a spot to stand and place a lever, he could move the world. Today's Archimedeans are still interested in leverage, but of the entrepreneurial type—gaining access to world markets and increasing trade. And one of the levers used by the modern Archimedeans is international standards.

Twenty years ago, companies wondered whether they were in a global industry and to what extent (if at all) they should have a global strategy. Today, companies have moved beyond this question and are trying to determine what their global strategy should be. Maturity of domestic markets has driven companies to pursue international expansion more aggressively. International trade agreements; multinational ownership or partnering of companies; the demise of the Eastern European communist bloc; the privatization of formerly state-owned enterprises throughout the world; the information technology revolution; the rise of newly industrializing countries; and the growing confluence of cultural tastes have all contributed to unprecedented global business opportunities.

It's a great time to be a global enterprise. But it is not a time without challenges. Trade barriers are falling, but they still abound. Sometimes trade barriers are in the form of tariffs. Sometimes quantitative restrictions, import licensing, or local content requirements can be barriers to trade. Conformance assessment can also create a trade problem. And sometimes, standards are the barrier.

At the Transatlantic Business Dialogue (TABD) meeting in November 1995, over 100 American and European business executives identified standards as one of the major barriers to trade. But as a general rule, these executives recognized that when standards are a barrier, it is a local, national, or regional standard that is the barrier. Rarely is the barrier an international standard. One of the key recommendations from the TABD was to eliminate trade barriers that result from restrictive standards by developing a common set of harmonized standards, preferably based on international standards.

Before we go further, international standards must be defined. Those standards issued by the International Organization for Standardization (ISO) and the International Electrotechnical Commission (IEC) are what many people think of in terms of international standards. There are also intergovernmental organizations that produce international standards, such as the International Telecommunications Union (ITU) and the International Organization of Legal Metrology (OIML). But the true test for whether a standard is "international" is whether it has broad recognition and acceptance around the world. Using this litmus test, the Boiler and Pressure Vessel Code of the American Society of Mechanical Engineers (ASME), the petroleum and plastics standards of the American Society for Testing and Materials (ASTM), the aerospace standards of the Society of Automotive Engineers (SAE) and Aerospace Industries Association (AIA), and others would also qualify as international standards.

INTERNATIONAL STANDARDS ARE STRATEGIC TRADE LEVERS

As part of their global strategy, most companies have embraced international standards as a key tool to open markets. Caterpillar, Hewlett-Packard, United Technologies Corporation, AMP, Ford Motor Company, and Unisys are just a few of the major corporations that foster the use of international standards in their strategic standardization-management plans. Twenty-five years ago, international standards only accounted for about 10% of the standards used by a company. Today, that figure is around 45% (1).

Why do companies place such emphasis on international standards? It's because international standards are an effective way to achieve the synergistic goals of tearing down trade barriers and creating global market opportunities. Of the $465 billion in U.S. exports in 1993, $300 billion were affected by non-

U.S. standards (1). Would a U.S. company prefer those non-U.S. standards to be international or non-U.S. national standards? Would a U.S. company prefer to comply with a single international standard or many different national standards? These rhetorical questions are important because the answers affect global market accessibility and billions of export dollars.

The top ten U.S. export industry sectors, which include aerospace, automotive, telecommunications, plastics, and petroleum, have heavy U.S. participation in the development of international standards. These types of advanced-technology products are directly influenced by international standards, and they accounted for a U.S. trade surplus of $25.8 billion in 1993. In contrast, there was a trade deficit of $141.6 billion in those areas where either there are few international standards or an absence of U.S. participation in the development of the international standards (2). Obviously, the factors that affect trade balances are numerous and complex and many of them have little to do with standards. Nevertheless, the correlation between international standards, U.S. participation, and national trade surpluses seems to be more than a coincidence.

Companies use international standards as strategic levers primarily in four ways to topple trade barriers or pry open doors to increase existing or create new trade opportunities. First, international standards provide companies and their governments with the means to challenge national standards as being restrictive trade barriers. Second, international standards provide business opportunities by unifying technical requirements, thereby unifying markets. Third, there is a strong link between global trade and global production. International standards allow companies to produce and market the same product globally. Finally, the rapidly changing political, social, and economic landscape has opened many new markets around the world. But emerging trading partners find themselves in need of standards, and international standards are helpful for market entry.

COUNTERWEIGHT TO PROTECTIVE NATIONAL STANDARDS

Trade experts estimate that U.S. companies could export an additional $20 to $40 billion a year if it were not for technical barriers to trade. The most significant technical barrier to trade is the result of differences between U.S. standards and conformity assessment practices and those of our trading partners (3). In some cases, it is an international standard that creates the technical barrier to trade. This usually happens when the United States does not participate in the development of the international standard. But in most cases, it is a national or regional standard that creates the barrier, which is a situation where the United States has limited or no opportunity to participate in the development of the standard.

In the absence of an international standard, it is difficult (though certainly not impossible) for the United States to persuade a trading partner to revise or remove their restrictive standard. Without an international standard, there is no unbiased arbiter to settle technical disagreements. It becomes a matter of your standard versus my standard; your practice versus my practice. But with an international standard, arguments for maintaining a restrictive national standard begin to crumble, and pressure can be brought to bear for changing the restrictive national standard lest a nation be found guilty of violating the General Agreement on Tariffs and Trade (GATT).

When Poland issued a regulation requiring many imported products to obtain safety certification from its Center for Testing and Certification (PCBC) or one of the fifteen institutes supervised by the PCBC, this regulation presented a trade barrier. Aside from the additional cost of testing, Polish standards and certification practices differed from international ones. In some cases, this meant that U.S. products, which complied with international standards, could not be imported because they did not meet the restrictive Polish standards.

As a result of efforts by the U.S. and the European Union, Poland is changing its product standards and will allow for producer or third-party certifications to confirm compliance with the appropriate international product standards. Poland expects to have their regulations changed in early 1998, and has agreed to suspend the requirement for PCBC safety certification until January 1, 1998 (4). If not for existing international standards, resolution of this dispute would have been more difficult, and Polish markets closed to some U.S. products.

UNIFY WORLD MARKET REQUIREMENTS

Oscar Wilde wrote, ''When the gods wish to punish us, they answer our prayers.'' For many years, some U.S. companies wished for a single set of standards in Western Europe. Companies had or still have to produce as many as nineteen different versions of their products due to differing national standards. In the 1990s, the gods granted the wish of those seeking unified European standards when the European Community aggressively developed numerous standards. The only problem was that if these standards were not in harmony with the U.S. standards, companies either had to make expensive changes to their production lines, abandon the lucrative Western European market, or try to convince the European Community to change their restrictive standard.

Recent journal articles and government reports are replete with examples of U.S. companies marketing safe, efficient products in some European countries one day, only to find themselves without a market the next day when a unified European standard changes the requirements. For example, Zimmer & Company, the world's largest producer of orthopedic devices, had to spend $5 million in testing equipment to make sure that a socket ball in its hip replacement line was smoother and rounder in order to conform to an arbitrary detail requirement in a European standard. There had never been a complaint before about the smoothness and roundness of Zimmer's hip replacements, but in the end Zimmer had to invest the money or risk losing its substantial European market. Caterpillar had to make similar investments to install sound-suppression devices on some of its earth-moving equipment transmissions, even though the other major world markets did not require it and there was no customer demand for such devices (5).

Zimmer and Caterpillar coped with the new regional European standards because they had a significant market presence and the money to make changes to their products. But what of small firms that cannot afford the changes, or companies whose European market presence is marginal? They can spend years trying to fight the unfair trading practice through official government channels. They can try to gain a nonvoting seat on a European standards developing committee and influence the requirements. These tactics have some examples of success, but usually with a significant loss of time and revenues. The best tactic, however, is the proactive one. Preempt potentially restrictive national and regional standards efforts by initiating international standards activity or the adoption of a U.S. national standard as the international standard. Such efforts do not guarantee success, but the international forum offers a much more level playing field.

The Association for the Advancement of Medical Instrumentation (AAMI) has been very proactive in participating in the development of international standards to ensure continued U.S. leadership and competitiveness in the medical-devices global markets. When the European Community began developing unified medical-device standards in the late 1980s, AAMI moved quickly to ensure that the United States had a strong voice in the ISO and IEC committees, as well as access to the European standards committees.

For example, in 1990, AAMI was instrumental in forming ISO technical committee 198 on sterilization and providing the secretariat. Because of AAMI's quick action, the international sterilization-equipment standards preempted unified European standards, which may have excluded U.S. products. There are unified European standards in this area, but they are harmonized with the international standards (6).

Before 1990, the exports of Amsco, a leading U.S. producer of steam sterilizers, went to Canada, Saudi Arabia, Singapore, and Taiwan—places where the national standards were compatible with U.S. standards. European sales were slight because the company could not justify the cost of building a wide variety of different models to satisfy the national standards of each European nation. Amsco welcomed the effort to develop international standards as a way of gaining increased market entry into Europe and elsewhere. Amsco funded participation on both the European and ISO standards committees as an investment in their future (7).

In 1996, U.S. medical technology exports reached $12.9 billion, with a record $6.5 billion trade surplus. Of these exports, $5.1 billion went to Europe, and the U.S. trade surplus with Europe was $3 billion (8). We can only hypothesize whether this would have been the situation in the absence of harmonized international standards.

PROMOTE GLOBAL PRODUCTION

In simpler times, production took place in a single country, by a single manufacturer, using local suppliers. Today, multinational corporations abound. Former competitors are now partners in strategic alliances. Companies continue to acquire interests in foreign firms to provide easier access to markets. Manufacturers now search globally instead of locally for the best suppliers. Products are becoming increasingly designed, manufactured, and supported on a worldwide basis, taking advantage of the different talents around the world, while reducing the cost of labor, materials, and transportation. We are entering an era where no single company or single country can successfully compete alone in the new global game.

The automotive industry provides the best example of widespread global production. In recent years, Ford has acquired an interest in Mazda; Chrysler in Mitsubishi; and General Motors in Isuzu and Suzuki. General Motors and Toyota built a joint manufacturing facility in California. Ford, General Motors, and Chrysler work together under a joint venture called the Partnership for a New Generation Vehicle to ensure the future competitiveness of the U.S. auto industry. The Mazda Miata can be referred to as a "world car" since it was designed in California, assembled in Michigan and Mexico, and some of its electronic components were invented in New Jersey but manufactured in Japan. The increasingly global nature of the automotive industry has also increased the demand for international standards, which can be used to facilitate coproduction efforts, reduce costs, and eliminate differing regulatory standards that have become a major trade barrier.

When over 100 top executives from major U.S. and European companies met at the TABD in Seville, Spain in 1995, they specifically cited the need for harmonization of motor vehicle standards. While the automotive industry has more global production than most other industry sectors, the lack of internationally harmonized standards reduces the economic advantages of free trade and fragments prospective markets for products that could otherwise be produced identically at plants around the world. Having to produce right-hand drive cars for the United Kingdom and left-hand drive cars for the United States and continental Europe is one of the most frequently cited examples of a differing technical standard that forces manufacturers selling to multiple markets to produce different versions of the same product.

The TABD identified vehicle-safety requirements and environmental emissions as two areas in special need of harmonized international standards since these areas are frequently subject to government regulatory standards that have become major barriers to trade and cost effective production. In fact, these different regulatory and certification requirements may add more than 10% to the design and development costs of a vehicle (9). For the manufacturer, the increased costs and design risks associated with producing a different model for each market may be enough to prevent or limit market entry. For the consumer, it certainly means higher prices and reduced choices.

As a result of a TABD recommendation, the European Institute hosted the first Transatlantic Automotive Industry Conference on International Regulatory Harmonization in 1996. The conference included a broad cross section of the automotive industry, including the American Automobile Manufacturers Association and the Association of European Automobile Constructeurs, as well as government officials from regulatory agencies. Their goal was straightforward: completely eliminate trade barriers resulting from unwarranted differences in vehicle standards and certification procedures. The task of accomplishing the goal, however, was complex because it involved harmonizing many different automotive standards.

The conference concluded with a number of action items. Some of the actions were as simple as seeking worldwide acceptance of the noise test procedures in ISO 362; while efforts to harmonize fuels and emissions standards are more of a challenge and will take longer. Both U.S. and European Union regulators have committed to changing regulations, but the key rests with industry uniting behind harmonized international standards. Given the potential gain in global market accessibility, manufacturing efficiency, and increased profits, the chances of success seem good.

TO GO WHERE FEW INTERNATIONAL STANDARDS HAVE GONE BEFORE

The radically changing world political scene has created many new market opportunities. The dissolution of the Soviet Union has resulted in many new independent states. Deregulation and privatization of formerly nationalized industries is occurring at a rapid pace throughout Latin America and parts of Asia and Africa. Such Pacific Rim nations as Indonesia and Malaysia are emerging as the next generation of industrial tigers that are eager for more trade. And as China gradually lessens its political and economic restrictions, the largest trade opportunity of all is opening.

One feature all of these emerging trading partners share is either a lack of standards; a system of standards previously or currently directed by a state bureaucracy motivated more by centralized control than by world trade; or a system of standards founded on obsolete, parochial, or unnecessarily restrictive processes. These nations are not only looking for standards that will ensure the quality, safety, and environmental suitability of products imported into their countries, but reciprocal standards that will allow them to export to the rest of the world as well.

At a recent meeting of the Asia-Pacific Economic Cooperation Council (APEC) countries, the Subcommittee on Standards Conformance (SCSC) discouraged member nations from developing internal product standards since these often tended to contribute to trade barriers. Instead, the SCSC recommended its members either accept existing international standards or realign their product standards with those standards accepted worldwide (10). The emerging industrial nations in particular understood that trade is a major engine of economic growth, and that international standards do a better job at fueling that engine than do national standards. They also appreciated that even though they expose their fledgling industries to some risk when transitioning from protective internal standards to international ones, domestic industries will ultimately benefit from the efficiencies that international competition and cooperation foster.

The recommendations of the SCSC were not new, but part of a larger effort to increase free trade among the Pacific Rim nations that began with an APEC summit in 1993. Are the APEC's efforts to emphasize international standards working? Like anything else, success is mixed, but there are heartening examples. The National Standardization Council of Indonesia (Dewan Standardisasi Nasional, DSN) is committed to ensuring that the existing 3,550 Indonesian standards are in harmony with worldwide practices. Following recently changed processes, the DSN first determines if an ISO or IEC standard exists before developing or revising its own standards. If such international standards do not exist, the DSN seeks national standards that have the reputation of being used worldwide. For example, the new Indonesian cement standards are based on ASTM standards that are de facto international standards (11).

The U.S. Department of Commerce considers Indonesia to be one of the top ten emerging markets where U.S. exports have enjoyed a growth rate of nearly 20% a year from 1988 to 1997. The reason for such an increase is the continual liberalization of Indonesian trade policies, part of which is based on the migration from national standards to worldwide standards.

MORE ON STANDARDS AND TRADE

While I have focused on the four primary universal reasons international standards help build trade, there are many other ways in which international standards contribute to trade on a more individualized basis. For example, when Canon first entered the photocopier market in the early 1970s, it decided from the outset to design a global product. In contrast to its Japanese competitors, Canon chose to design a copier using ISO and U.S. paper sizes instead of trying to develop multiple versions of the same product to accommodate unique Japanese paper sizes as well. The presence of international standards gave Canon the opportunity to make a strategic business decision to ignore some of its domestic market in order to maximize its overall profits and future prospects in the global market (12).

We usually think of trade in terms of commerce between nations, but there is also the trade between nations when cooperating in military operations. Interoperability is key to any successful joint military effort, and interoperability has been a central strategy for the member nations of the North Atlantic Treaty Organization (NATO) since its beginning. In the past, this interoperability was largely achieved through the use of military standards. But as defense budgets are reduced throughout NATO, the member nations are placing more emphasis on ISO or IEC standards in order to control costs by buying commercially available products, ensure adequate supplies in times of war, and optimize interoperability. For example, all NATO tactical shelter sizes must conform to ISO standards to ensure transportability on European rails and highways. ISO standards also form the core of most NATO information technology requirements. In the absence of such standards, U.S. or European standards that are de facto international standards are used. For example, NATO automotive gasoline conforms to ASTM D 4814 or European Committee for Standardization (CEN) standard EN 228, and NATO kerosene conforms to ASTM D 3699.

ALL THAT GLITTERS IS NOT GOLD

King Hiero II of Syracuse knew that all that glitters is not gold, which is precisely why he tasked Archimedes to devise a standard test method to determine the gold content in his crown. It should not be assumed that there is an inherent goodness in a standard just because it carries the international label. The jury is still out on whether horizontal international standards, such as the ISO 9000 quality standards or ISO 14000 environmental management standards, which apply across all product lines, will ultimately prove to be an overall benefit or just an additional cost. For some companies, the cost of certifying to these standards may keep them out of the global market.

U.S. companies must also remain vigilant about the development of international standards and involved in the process. While international standards committees provide a forum to influence standards development, if companies don't participate, the resulting international standard may exclude them from world markets or increase production costs. For example, when an ISO standard for fire-sprinkler metal fittings was being developed, the design requirements were based on the fittings of several European manufacturers. If approved, the ISO standard would have excluded a stronger fitting made by a U.S. manufacturer, the Grinnell Corporation. Grinnell was alerted in time to lobby successfully for a change to the proposed standard, but if they had not, they would have had to abandon some lucrative world markets or spend hundreds of millions of dollars for retooling (13). While international standards tend to be less restrictive than most other types of standards, any standard can become a trade barrier.

"EUREKA!"—INTERNATIONAL STANDARDS AND BARRIER-FREE TRADE

Barrier-free trade is going to happen. It will happen in ebbs and flows. There will be the inevitable setbacks and horror stories. But in the end, the pragmatism of free trade will overcome the parochialism of protectionism. The only questions are how soon we will arrive at a barrier-free trade world? What will be the cost? How difficult will it be? Harmonized international standards can make the answers to these questions, ''sooner, cheaper, and easier.''

We keep looking for complex solutions to complex trade-barrier issues, when international standards hold many of the answers. When Senator Orrin Hatch introduced legislation in 1990 to recognize National Standards Week, he noted that, ''we have spent millions of dollars promoting U.S. products overseas and engaged in expensive negotiations with our major trading partners to try to open foreign markets to American products. However, I believe that we have overlooked one of the simplest and least expensive methods of making our products more competitive. I am talking about effort to promote harmonized product standards.'' Senator Hatch could not have been more correct.

International standards may not be a panacea in building trade. But the examples of where international standards have torn down barriers and opened trade doors far outweigh those examples where international standards have been a trade barrier. Moreover, the statements and actions by world leaders in government and industry suggest that they are committed to developing and using harmonized international standards to enhance trade opportunities.

When Archimedes finally discovered a way to determine the purity of the gold in his employer's crown while taking his afternoon bath, he reportedly jumped from his bath running through the streets of Syracuse shouting, ''Eureka,'' which is Greek for ''I have found it.'' If in the future, a coworker runs down the halls of the company shouting ''Eureka,'' don't be alarmed. Chances are it's just a modern day Archimedean who has discovered an international standard that's good as gold. And that means more leverage for your company in the global market.

REFERENCES

1. National Research Council. Standards, Conformity Assessment, and Trade into the 21ˢᵗ Century. Washington, DC: National Academy Press, 1995, p. 111.
2. National Research Council. Standards, Conformity Assessment, and Trade into the 21ˢᵗ Century. Washington, DC: National Academy Press, 1995, pp. 47 and 105.
3. KD Gasserud. NIST Standards in Trade Program. National Institute of Standards and Technology. October 29: 1, 1996.
4. Office of the United States Trade Representative (USTR). 1997 National Trade Estimate Report on Foreign Trade Barriers. p. 332.
5. P Oster. Europe's Standards Blitz Has Firms Scrambling. The Washington Post October 18: H1 and H4, 1992.
6. Defense Systems Management College. Standards and Trade in the 1990s: A Source Book for Department of Defense Acquisition and Standardization Management and their Industrial Counterparts. no date, pp. 3–12 and 3–13.
7. J Burgess. Competing in a Diverse Market. The Washington Post December 2: A1 and A6, 1991.
8. News Release from the Health Industry Manufacturers Association (HIMA). International Agreement Eliminates Redundancy, Cost, Helps Modernize the U.S. Food and Drug Administration. May 28, 1997.
9. International Trade Agency. Overall Conclusions of the Transatlantic Automotive Industry Conference on International Regulatory Harmonization. April 11, 1996.
10. Philippine News Agency. SCSC Pushes International Product Standards Harmonization. Manila Press Review October 21, 1996. APEC consists of Australia, Brunei, Canada, Chile, China, Indonesia, Japan, South Korea, Malaysia, New Zealand, Papua New Guinea, the Philippines, Singapore, Taiwan, Thailand, and the United States.
11. National Research Council. Standards, Conformity Assessment, and Trade into the 21st Century. Washington, DC: National Academy Press, 1995, p. 143.
12. GS Yip. Total Global Strategy, Managing for Worldwide Competitive Advantage. Englewood Cliffs, NJ: Prentice Hall, p. 93.
13. P Oster, V Gagetta, R Hoff. 10,000 New EC Rules. Business Week September 7: 50, 1992.

16

Standardization and Technical Trade Barriers: A Case in Europe

This case study on technical trade barriers and one demonstrable U.S. experience in Europe was contributed by Helen Delaney while she was a standards and trade specialist working with the U.S. Commercial Service in Brussels, Belgium. Prior to that she was ASTM's director of global affairs and general manager of its Washington DC office. She is currently President of Delaney Consulting in Bethesda, MD.

> *There was a little girl,*
> *Who had a little curl,*
> *Right in the middle of her forehead;*
> *And when she was good*
> *She was very, very good,*
> *But when she was bad she was horrid.*
> *—Henry W. Longfellow*

Throughout a relatively long age of innocence, standardization was looked upon as the hallmark of a civilized society. It was a temperate, consensual activity whose end product was an advancement in and of itself. A standard was an invention: a language, a technique, a plan drawn by the skilled and expert, a whole greater than its parts.

The use of standards in any society was directly related to its quality of life; and while its citizens readily accepted the fruits of this noble labor, the occupation itself was seen in turns as either esoteric or drab. It was left, therefore, to the technicians. However, progress and a burgeoning world market changed everything.

After the General Agreement on Tariffs and Trades (GATT) (1) and other trade liberalization efforts that removed artificial barriers, markets eased open around the world. Forced to defend home turfs against international rivals, manufacturers had to produce better and more affordable goods. Many were forced to seek markets offshore. The more ardent among them also looked for a substitute to fill the gap left by the reduction of tariffs. They found it in the hallowed but tedious halls of standardization.

Now, standards are all they used to be, but much more. The interest in standardization is sharper, more universal, more focussed in its effects on international trade and competition. Standards are now, according to the European Commission's Green Paper (2), "too important to be the exclusive preserve of technical experts."

When they are good, standards maintain a free flow of trade while rationalizing and interpreting laws or practices that protect a country's citizens from products that harm or defraud. When standards, either by chance or by choice, bar products unnecessarily from a market, when they are designed to protect industrial cabals, when used to distort trade in ways unrelated to health or safety or the environment, standards and the havoc they can wreak on industries are serious impediments. They are then referred to euphemistically as technical barriers to trade.

Nowhere was the elimination of technical barriers to trade more important than in Europe, where the integration of an economic space was creating the European Union (EU) (3). Few things were more detrimental to the formation of an internal market than separate, diverse, or conflicting standards. Few environments were more conducive to the creation of technical barriers than separate, diverse, or competing groups of standards developers. The solution was an integrated, centralized standardization system. How ironic it is, then, to find technical barriers to trade produced by a system that was designed to eliminate them! It is not so unnatural, however, when one realizes that the overriding principle in the creation of EU was a Europe without borders, a Europe where goods and people would flow freely, a Europe without technical barriers. A Europe for Europe first and others second.

While the following is an account of the experiences of the Dormont Manufacturing Company and its President, Mr. Evan Segal, and what he and others perceive as the formulation of a technical barrier to trade in Europe, the reader should keep in mind that technical barriers can happen wherever there is standardization.

The Dormont Manufacturing Company once sold its products all over Europe. The company manufactures hoses that connect gas appliances to gas outlets. Its troubles began in 1989 when its customer, Frymaster, was told by McDonald's it could no longer use the Dormont hose in British restaurants. Soon after, Euro-

Disney (4) followed suit and Dormont hoses were removed from establishments in the United Kingdom and France. The reasons given for both actions was that the Dormont hose was not designed to meet United Kingdom or French authorities' standards and safety rules.

The European Commission has made great efforts to bring the individual safety laws of its member states into homogeneous, or "harmonized" Europe-wide directives. Many products, however, remain regulated under national laws. Among them are gas connector hoses. The same is true of standards. Comite Europeen de Normalisation (European Committee for Standardization, CEN), the European regional standardizing body based in Brussels (whose membership consists of all member state national-standards bodies plus the bodies of several other European and Scandinavian countries) has developed European-wide standards for many products. As of this printing, however, no European standard exists for this particular product. Gas connector hoses must, therefore, comply with individual national member state standards.

Dormont decided to have its hose tested to a British standard. The standard contained both design and performance elements; but BSI in the United Kingdom waived the design requirements. The British Standards Institution, assuming the hoses were covered by the European Gas Appliance Directive (5), issued a certificate for compliance with U.K. safety regulations for gas appliances. This allowed the company to affix the European CE mark to its hose.

In theory, EU operates on a mutual recognition system that allows a product that bears a CE mark issued in one member state to be sold in all member states. Dormont assumed it had won the right to circulate its product for use again in the United Kingdom and France and, in theory, in all of Europe. Its seemingly bright future darkened quickly, however.

Dormont's German competitor complained to the European Commission, claiming that hoses were not appliances, but components. The commission agreed that hoses were not appliances and, therefore, not covered under the Gas Appliance Directive. Therefore, it decreed, the British Standards Institution had made a mistake. The CE mark was ordered withdrawn.

The hoses were now officially subject to relevant national laws and national standards referenced by those laws. Most of the national standards contained design criteria that were based on European-manufactured products. Despite a perfect safety record, the Dormont hose was prohibited from use and barred from its U.K. market, its French market, and other member state markets because it didn't fit any of the design requirements.

The national standards of EU member states, for example, required the fittings and tubes of the gas connector hose to be made of 316 or 321 stainless steel or carbon steel with zinc cladding. Dormont hoses are made of 304 stainless steel. According to Evan J. Segal, President of Dormont,

''Most EU national standards for flexible gas pipes allow *only* 316 or 321 austenitic stainless steel. When the flexible hose is not annealed and welding is used to attached end fittings, this is appropriate. However, other grades (including 304) when fully annealed, meet all performance criteria. There exist relevant performance tests . . . that can be included in the standard. In fact, a truly performance based standard would not specify materials; it may suggest those that have been used, but would not limit the innovation of the manufacturer.''

In 1991, test reports from France on Dormont hoses showed that Dormont passed all performance requirements such as leakproofness and resistance to temperature, fire, and stress, but failed on design requirements for materials, welding, and flow rate, none of which, according to Mr. Segal, are related to national gas network differences or to safety.

The situation advanced. European manufacturers of gas connector hoses decided to develop a Europe-wide standard in CEN. A European standard replaces all national standards. The Dormont hose, to be sold anywhere in Europe, would have to meet the new standard.

Meanwhile, with the assistance of a U.K.-based representative, the U.S. Embassy in London, and The Department of Commerce Office of European and Regional Affairs, Mr. Segal successfully convinced the British Standards Institution's technical committee to revise the U.K. standard to reflect performance requirements. Dormont gas hoses were once again sold in the U.K. (only for commercial uses, however; domestic connectors are still barred, pending revision of the second part of the standard.)

While the European standard is being developed, however, all work on national standards on gas connector hoses must come to a ''standstill,'' a rule that would prohibit Mr. Segal from repeating this ''success'' in any other member state. At a meeting of the CEN working group, a working paper was put on the table by a German participant. It was design based.

Since gas connector hoses are not regulated at the European level, the CEN standard will be a ''purely'' voluntary one. However, it still has the potential to become a barrier to trade if compliance with it is required by architects, insurance carriers, banks, or other institutions with an economic or political interest, and/ or if the exporter is obliged to use the services of a third party to prove that his product conforms to it.

The European Commission has said it will consider whether or not it will amend the Gas Appliance Directive to include connecting hoses within the scope. All commission meetings to date, however, have resulted in no decision.

If there is a decision to include hoses, a Europe-wide law covering connecting hoses will launch an ''official'' European standardization procedure. The commission will mandate and find the development of a standard by CEN that will interpret the essential requirements of the directive (6). The commission will not intervene in the voluntary standardization process. If the hoses remain unregulated at the European level, the European standard will still be the standard to apply under member state authority. If it is a design standard that excludes the Dormont hose, the entire European market will be closed to this product, and it will once again be removed from the United Kingdom.

The U.S. Mission to the EU in Brussels appealed to CEN to permit Dormont to present its case to the Working Group (using the American National Standards Institute [ANSI] as the intermediary). The request was denied, although Dormont was allowed to submit a position paper to the Working Group. Under CEN rules, Dormont can also submit comments on a draft standard when it enters the public comment phase. The technical committee will be obliged to consider the comments and respond.

The CEN Secretariat (the managing body) has stated that its policy is in accordance with the World Trade Organization (WTO) Code of Good Practice (7), which calls for standardizing bodies to "ensure that standards are not prepared, adopted or applied with a view to, or with the effect of, creating unnecessary obstacles to international trade." The code also states that "wherever appropriate, the standardizing body shall specify standards based on product requirements in terms of performance rather than design or descriptive characteristics."

The United States Trade Representative (USTR) has raised the Dormont issue at meetings of the Technical Barriers to Trade Committee of the WTO, along with other similar standards issues that are developing in CEN. However, the time it will take to correct the situation may make the effort irrelevant to the U.S. company based in a town in Pennsylvania called, ironically enough, Export, PA.

Standards that are overly prescriptive, based on designs, specified materials, and other characteristics not related to safety or performance can have horrifying effects on companies and their employees. They also strain diplomatic, economic, and political relationships between trading partners.

Several factors are associated with the creation of standards barriers. First and foremost is the nature of "closed" standards organizations. The national standards bodies of practically every nation in the world, including the standards organizations of the EU, have a closed-door membership policy. With the exception of Japan, the United States may be the only country in the world where standards organizations open their membership to direct international participation. (Since language is a distinct barrier for Westerners, few take advantage of the opportunity to participate in Japan). If Dormont had a facility in Europe, Mr. Segal would have a better chance of participating in the CEN process, qualifying as a European.

Most standards bodies in industrialized nations are private-sector organizations and are not obligated to honor the terms of the WTO Technical Barriers to Trade Agreement, which is binding only on governments. Those that do ascribe to the Code of Good Practice do so voluntarily, but without enforcement mechanisms. The secretariats of such organizations, while claiming adherence to WTO principles, also maintain a contrasting ''hands off'' policy with their technical committees and technical experts, leaving their autonomy relatively intact.

Many argue that standards developed in the private sector are voluntary, and that manufacturers have a choice of whether or not to use them. *Voluntary*, however, is a relative term in the real world. There is nothing voluntary about a standard that will not be honored by market forces like insurance companies, architects, banks, or retail merchandisers.

Besides multilateral institutions and Section 301 of the Trade Act of 1974 (8) as amended, and the adherent long-term processes of negotiation and dispute settlement, this leaves the immediate business of barriers to trade to those who create them.

Until the creators realize that those who restrict are themselves restricted, that barriers created to protect also imprison, and that trade is best served when technology is allowed to advance in an open environment, technical barriers to trade will continue to disturb and distort markets everywhere.

REFERENCES AND NOTES

1. The GATT (1947) was directed to the substantial reduction of tariffs and other barriers to trade and to the elimination of discriminatory treatment in international commerce.
2. The Development of European Standardization: Action for Faster Technological Integration in Europe.
3. The European Union is comprised of the member states of Austria, Belgium, Denmark, Finland, France, Germany, Greece, Ireland, Italy, Luxembourg, the Netherlands, Portugal, Spain, Sweden, and the United Kingdom.
4. Now called Disneyland Paris.
5. 90/396/EEC. Official Journal 196: 15, June 26, 1990.
6. Although the ''official'' standard will be ''voluntary,'' it carries with it the presumption of conformity. Products not manufactured or tested to a European standard must be subjected to third-party certification, thereby increasing the cost of the product.
7. Annex 3 of the Agreement on Technical Barriers to Trade (Annex 1A of the Marrakesh Agreement Establishing the World Trade Organization following the Uruguay Round of Multilateral trade negotiations of the General Agreement on Tariffs and Trade).
8. Under Section 301, the USTR may request discussions for dispute settlement, and if the other parties refuse, the USTR may make recommendations to the President to proceed with retaliatory measures.

17

Safety Standards and Product Certification in a Global Market

The process of standards development and product certification are intertwined. This chapter explores current initiatives in standards harmonization, describes the U.S. and European product-certification processes, and provides some ideas for future initiatives. The chapter was prepared by Robert A. Williams, Corporate Manager, Standards and Research, for Underwriters Laboratories (UL) Inc.

THE SPECTRUM OF HARMONIZATION

The term *harmonization* refers to a process by which the technical requirements of various standards have been made equivalent or identical. Although the actual words comprising harmonized standards may be different, the performance requirements or safety issues have been addressed equally. By using a harmonized standard, manufacturers and certification agencies can be assured that any product tested to one standard will meet the requirements of another standard. Confusion exists over the meaning of the term *harmonization*. Some in the standards community believe that *harmonization* equals *identical*; others believe a standard is harmonized if the requirements are technically compatible. Additionally, harmonization with international standards is discussed quite often, although it is (generally unlikely or) virtually impossible to publish a U.S. standard that is identical to an international standard. Deviations are usually necessary. Though a quantifiable description of harmonization would be desirable, it would require detailed analysis and interpretation of requirements to arrive at such a description, and the degree of harmonization would probably remain debatable. We can define the levels of harmonization as:

Chapter 17

1. Identical standards;
2. Technically compatible standards; and
3. Standards adopted with deviations.

Further harmonization occurs through technical equivalency. Many times, requirements are different, (i.e., different methods used to conduct a test to evaluate the same property), but they both may establish a "level of safety" that is acceptably equivalent. Requirements that are technically equivalent are also technically compatible. For example, one standard may require an appliance to be subjected to a drop test and a corresponding standard may require an appliance to be subjected to an impact test. In each of these tests, the appliance is subject to impact. Different methods are used, but the end result of the tests could be considered technically equivalent in that the product has been tested for the property of impact resistance.

Standards Adopted with Deviations

Manufacturers trying to sell their products in foreign markets want to avoid delays and costs associated with compliance to multiple standards and gaining product approval or certification in each country. The country-by-country approach is time consuming and unnecessarily expensive, by contrast with compliance to regional or international standards.

The development of international standards is proceeding in two leading international organizations, the International Electrotechnical Commission (IEC) and the International Organization for Standardization (ISO). It's not a rapid process. Even slower is the adoption of such standards by most countries. Many industries are not ready to move to international standards. A number of UL standards have been harmonized, to varying degrees, with international standards because certain U.S. industries have expressed willingness to move to this direction.

Industry and user needs are the key to harmonization of UL and international standards. Harmonization with international standards may well require the involved industries to modify specific products; modifications that may require considerable investment of time and money. It becomes difficult for manufacturers to make changes that are costly, and add little or nothing to the safety of the products involved, simply for the sake of harmonization. And yet, as the motivation to seek foreign markets grows, conformity to international standards becomes a prerequisite. This trend has been promoted by European Common Market initiatives (EC1992), the US/Canada Free Trade Agreement, North American Free Trade Agreement (NAFTA), and other economic global developments.

Typically, when a testing laboratory or country adopts an international standard, the standard is adopted with deviations based on particular requirements for the adopting organization or country. These differences may be labeled either as "national conditions" or "national deviations." A national deviation occurs when regulations outside the control of the organization adopting the standard require an exception. Typically, a deviation is necessary because of the adopting country's installation, building, or other code that is not directly under the control of the adopting organization.

A textbook example of international standards harmonization occurred in the United States several years ago. UL and U.S. computer- and business-equipment industry devoted thousands of hours to develop UL Standard 1950, which is based on the international standard IEC 950. In this effort, conflicts were identified between IEC 950 and the National Electrical Code (NEC), as well as the National Fire Protection Association (NFPA) Standard 75. To eliminate these conflicts, proposals were then made to revise the NEC and the NFPA 75 Standard.

THE U.S. SAFETY SYSTEM

The national safety system in place in any country is a product of its culture. This is certainly the case in the United States, where the safety system is a unique mix of voluntary and mandatory components, with roles for a variety of interested parties. Safety in the United States is largely the result of three elements working together: (1) product standards, (2) installation codes, and (3) laws or regulations. While these elements are separate, they have two things in common. They identify the requirements to be met, and they contain mechanisms to demonstrate compliance to those requirements.

Product standards contain requirements for the safety and performance of products. Generally, standards are developed in the private sector by a variety of groups. Codes address the safe installation, application, and placement of products and materials. Like standards, codes are developed in the private sector. Significantly, these codes are generally adopted as law word for word or with minor modifications by local governments. The third element, national laws, mandate safety requirements covering employers, employees, and consumers.

THIRD-PARTY CERTIFICATION

There are a number of reasons why a manufacturer may decide to pursue third-party certification. In general, the chief reason is that market forces require evidence that products comply with standards, codes, and laws. Third-party certification can provide the bridge of confidence that a product complies with standards, codes, and laws.

As in any transaction, the seller wants to sell a product to a buyer. A buyer will only purchase the product if they have confidence that the product will fulfill a need. Along with usability, quality, service, etc., buyers need confidence that the product complies with standards, codes, and laws. Third-party certification can provide the bridge of confidence that a product complies with standards, codes, and laws.

Model codes typically reference standards. Model codes are often adopted as law in most U.S. cities and towns. The more than 40,000 jurisdictional authorities who are responsible for enforcing the codes need evidence that a product complies with the code. Certification of the product by a third party is recognized as evidence that the product complies with the code.

In the United States there are laws governing safety of products that must be complied with. These laws designate federal government agencies as responsible for specific aspects of product safety. For example, the Consumer Product Safety Commission has the power to force recalls of unsafe consumer products. The Federal Communications Commission (FCC) regulates the airwaves, while the Occupational Safety and Health Administration (OSHA) writes and enforces rules governing safety in the work place.

The architects and contractors who specify products, and the wholesalers, retailers, and other members of the distribution channels who move products to consumers require confidence that a product complies with standards. These distribution channel members understand that electrical and building inspectors look for third-party safety-certification marks. Further, these distribution channel members know that third-party certification can help to limit the risk that a consumer will be hurt by a product and bring a lawsuit against them.

Now, there is a new term, *globalization*. This concept of globalization has gained tremendous momentum in recent years. There is hardly an industry meeting that does not include seminars, meetings, and discussions relating to the expansion of foreign markets, foreign challenges to the U.S. domestic market, concerns about barriers to trade, impact of fluctuations in the value of the U.S. dollar, and competitive pressures brought on by strides in foreign technology. The concept of a single world market and its challenge to the United States and Canada is equivalent in impact with other pressing problems facing these governments, such as the deficit, imbalance of trade, and threat to leadership in technology.

BILATERAL RECIPROCAL ARRANGEMENTS

A related imperative to standard harmonization must be treated—namely, some form of reciprocity of test data or certification among product-approval systems in various countries. In this case, the process becomes even more complicated than harmonizing standards. There are major differences in product approval or certification requirements in different countries ranging from practically no approval requirements in the emerging countries to some legislated requirements in many countries to the fairly sophisticated national, state, and local enforcement system used in the United States; and the national and provincial system in place in Canada.

In recognition and support of the objective of manufacturers to reduce the cost and delays in achieving product safety certifications in different countries, UL has established bilateral reciprocal arrangements with a number of organizations in Europe, Japan, and Canada. These arrangements have been methodically established for specific products, with cross training, round-robin testing, mutual witnessing of tests, and other measures to assure duplication of results. Establishing bilateral reciprocal arrangements is expensive and is undertaken only for industries that can demonstrate and document the need for such reciprocal arrangements. These bilateral reciprocal arrangements may be altered in the future. Decisions being made, and which will be made at the national government level, may well change the approach.

As in all international negotiations, there is a quid pro quo. That is, gaining acceptance of UL test results on products of U.S. manufacturers in other countries means a willingness to accept test results from laboratories in other countries on products from their manufacturers. Until 10 or 15 years ago, conformity assessment was carried out by official organizations, in the majority of cases only one organization per country, often directly authorized by the state, and sometimes operating in a regulatory context that made certification compulsory. In any case, such organizations operated in a consensus environment and all relevant interests at the national level were represented in the certification organization. Furthermore, each certification organization granted its own mark of conformity or a national mark owned by the state. The meaning of such a mark was conformity to the national standards.

In recent years, this situation has changed. Compulsory certification, where existing, was eliminated in most countries and certification is more and more seen as a service required by the market (by manufacturers, by consumers, and, sometimes, by authorities having jurisdiction or AHJs); and as such, as a service to be granted under free competition rules. Until 20 years ago, each country had its own set of standards that differed from each other. A product to be marketed in various countries had to meet all the different standards of each country. The increase of international trade made it necessary to harmonize these varying standards that were otherwise considered a technical barrier to trade.

The consequence of the harmonization process on conformity assessment is that the meaning of all marks of conformity tend to be the same, in other words, conformity to the international standards. This is a great change; all the

different marks, even with different shape, mean the same. This fact cannot be
ignored and from a commercial point of view may have some consequences.
When a product is marketed in various countries, it has the need that the
requirements to comply with are the same (need of harmonization of standards)
and the need not to repeat unuseful tests in the various countries. As a consequence, these issues have given rise to various agreements among certification
organizations within economically homogeneous areas. Due to the IECEE/CB
scheme inherited from a preceding European scheme, which now has spread
worldwide, the mechanism of the mutual recognition of test results is growing.
The great development of the CB scheme (more than 6,000 CB test certificates
issued in one year) demonstrates the effectiveness of this mechanism as a response to the real trade issues of manufacturers. See also Chapter 9 for details.

NEED FOR AN INTERNATIONAL MARK

Previously, when manufacturers made use of mutual recognition agreements on
test results of the CB scheme, the final certification granted was always one or
more marks of conformity owned by each certification organization involved.
With these agreements, certification is represented by one mark of conformity
having a regional or wider recognition. The same mark may be granted by anyone
of the organizations participating to the relevant agreement. Examples of this are
some European marks. A foreseeable next step could be an international mark
having a worldwide acceptance. Four main conditions must exist for implementation of an international mark of conformity:

1. The standards used for assessment must be equal in all countries participating in the scheme.
2. The participating countries must be economically homogeneous.
3. The program must have the support of the relevant affected industries.
4. Participating certification organizations must cooperate.

Furthermore, all certification organizations participating in the scheme must have
a well-established certification scheme in their own country and must be committed to recognizing the common mark as equivalent to their own mark. A consequence of this trend from nationally recognized marks to an international mark
means better service for industry. But it also means a further reduction of business
for certification organizations, since type test, initial assessment, certification,
and follow-up inspections will be conducted by a single organization. The advantage for industry is that they deal with a single certification organization for certification (registration or accreditation) recognized in several countries.

18

Consumer Product Safety in the Global Marketplace

This chapter is contributed by the Honorable Nancy Harvey Steorts, former Chairman, United States Consumer Product Safety Commission (CPSC) and former President, Dallas Citizens Council. She is President, Nancy Harvey Steorts International, Chevy Chase, MD and served as a Director of the American National Standards Institute (ANSI) and as Chairman of ANSI's Consumer Interest Council (CIC). She has also served as head of the ANSI/USA delegation to the International Organization for Standardization's (ISO) Consumer Policy Committee (COPOLCO). This paper was presented at the Pacific Area Standards Congress XX in April 1997 and has been edited especially for this book.

BACKGROUND

It is very important to have chosen the topic "Consumer Safety in the Global Marketplace" for a major address at the 20th Pacific Area Standards Congress (PASC), as it is so very important. Toys, car seats, pacifiers, foods, air bags, children's products, chenille sweaters, skirts, carbon monoxide detectors, electrical appliances, and sporting equipment are just a sampling of some of the products that are causing concern for consumers today. As the former Chairman of the U.S. CPSC, I have spent many years working with government leaders, industry leaders, standards officials, and consumers in an effort to make the products in our marketplace as safe as possible for all consumers. The consumer product-safety issues are an emerging trend that the nations of the Pacific area and the rest of the world are beginning to recognize as being an extremely important concern.

THE INTERNATIONAL MARKETPLACE

In the last few years, the world has undergone profound change. With the collapse of socialism, new markets have opened and new, resurgent economies, such as in the Philippines, Malaysia, Singapore, Thailand, China, Japan, and Korea, show remarkable vitality, energy, and power. Corporations are rushing to take advantage of fantastic opportunities that few people had envisioned just a few years ago. With these opportunities come competition and other obstacles that businesses must overcome in order to thrive. The emergence of the Far East economies has fundamentally shifted the economic balance of power and creates a new source of competition in the global economy. The Far East, as well as the European Union, Japan, and the United States, are major players in this new marketplace. Almost all countries throughout the world are now beginning to emerge economically. It is important that as they develop products that will be sold to their consumers, as well as to consumers in other countries, that they be safe, of good quality, and reasonably priced.

Pascal Krieger for ISO Bulletin, reprinted with permission from ISO, International Organization for Standardization, Geneva, Switzerland.

The consumer throughout the world today is changing. Buying patterns are changing. Higher expectations for quality products and services are being demanded throughout the world. The explosion of information technology has made consumers, government leaders, standard officials, and industry leaders much more aware of what is happening, from one country to another within a matter of hours. No longer is it acceptable for a product that is unacceptable in one country, to be sent to another country without notification and permission.

WHAT IS THE ATMOSPHERE IN THE UNITED STATES TODAY?

The American consumer is demanding. The consumer expects quality and safety in the products in which they invest their money and their trust. Consumers expect nothing more than to have total confidence in their purchases. If products are not up to their expectations, they expect something to be done about it. Safe, quality products are the rule. Furthermore, this rule cannot be violated. For a business to succeed in the U.S. marketplace, this fact must be realized. For businesses that do not heed this warning, punishment is severe. The work of the U.S. CPSC and the Food and Drug Administration (FDA) has raised the consciousness of the public. When the media began to report hazardous products such as toys, clothing, food, drugs, cosmetics, and housewares; new attitudes adamantly stated that not only could something be done to reduce hazards, but that it must be done. As a result of this public pressure, government regulatory agencies have been given many options to counter products that violate U.S. product-safety standards. Litigation and product recalls are two of the most formidable actions that can be taken. Safety is assured by the government's reliance upon vigilant and vocal consumers who expect quality at a fair and reasonable cost.

Coincidental with a rise in safety consciousness came an increase in litigation over product liability. Juries began to award large damages—actual and punitive for death and injuries caused by products. The courts have stated unequivocally that such suffering is preventable and must be eliminated. Consumers are reporting problems and corporations will be held responsible. Finally, over the years, consumer groups have formed that monitor product safety and report concerns to the media and to the government.

The U.S. Consumer

The fact that the U.S. economy and society are changing can be verified by observing today's retail market. Such changes reflect or can be caused by the changing habits of the consumers. Consumers do not behave the way they did in the recent past. They are smarter, less materialistic, more demanding, possess less disposable cash, are very diverse, increasingly serious, and financially insecure. They expect excellent quality and service at a low cost. These trends will continue in the near future.

Consumer-driven Economy

In the future, successful marketing and advertising will begin with the understanding that the economy is consumer driven. Companies must contend with the new reality: supply exceeds demand. Evident in the current barrage of price wars, coupons, promotions, and other forms of discounting by retailers; a new emphasis on safe products, quality, and service has been spurred. This is what consumers demand, and when consumers demand, the marketplace must provide.

External Influences on Consumers

Consumers are under great pressure from the globalization of the economy, economic restructuring, the information revolution, and stagnant wages. All of these things combine to create the potential for great anxiety and insecurity within the consumer. These "macroeconomic" pressures combine with social trends to create a much more complex marketplace, that international business must understand in order to survive and remain profitable. Social trends include the redefinition of the family, the rebirth of social activism, the focus on personal and environmental health, a lessening allegiance to work for self-fulfillment, increasing self-reliance, and the emergence of the "Professional Consumer." These social trends mean that consumers are becoming harder to understand and much more complex. Consumers are redefining their priorities. While working still matters a great deal to consumers, they are also placing greater importance on their families, their health, and the environment. It is up to us—the international product-safety community—to assist the consumer and assure that dangerous products not only never reach the marketplace, but are not manufactured in the first place.

Consumer Bill of Rights

In 1994 in the United States, President Clinton proclaimed the right of consumers to safe products, quality products, and good service. These consumer rights were first articulated by President Kennedy, on March 15, 1962, in a Special Message to Congress on Protecting the Consumer Interest. The Consumer Bill of Rights includes:

1. *The right to safety*: The right to expect that the consumer's health, safety, and financial security will be effectively protected in the marketplace.
2. *The right to information*: The right to have full and accurate information upon which to make free and considered decisions, and to be protected against false or misleading claims.
3. *The right to choice*: The right to make an informed choice among products and services in a free market, at fair and competitive prices.
4. *The right to consumer action*: The right to a full and fair hearing, and an equitable resolution of consumer problems.
5. *The right to consumer education*: The right to education, without which the consumer cannot enjoy the full benefits of the other rights.
6. *The right to service*: The right to convenience, courtesy, responsiveness to consumer problems and needs, and all steps necessary to assure that products and services meet the quality and performance levels claimed for them.

The Business Perspective

It is essential that all domestic and international manufacturers, who sell products in the American marketplace, be conscious of the consumer's vigilance. For the business that is not responsive to the consumer, litigation and other penalties await, such as lower profits and bad public relations. Successful businesses must realize this fact, otherwise, punishment is severe. The challenge for business, the government, and the standards community is not just safeguarding the domestic marketplace, but adapting to a market that is quickly changing due to international trade and increased consumer awareness. Superior quality and safety must be the highest priority. Real change requires a long-term, deep commitment and is becoming crucial to compete in the international marketplace. The United States must be as vigilant as we expect other countries to be when we sell our products in the international marketplace. "Made in the USA" stands for quality, therefore only quality goods should be exported to our colleagues around the world. There can be no dumping of products that do not meet our standards in other countries unless the country is made aware of the potential problem prior to shipment and consents to the shipment.

U.S. Consumer Product Safety Commission

Companies who do business in the U.S. marketplace must be familiar with the U.S. CPSC. This agency possesses broad power to ensure the safety of a wide range of products in the American marketplace. Toys, outdoor equipment, in-line skates, bicycles, indoor air quality, smoke detectors, flammable products, electronics, electrical wiring, furniture, textiles, housewares, fireworks, computers, clothing, and many other products fall within the jurisdiction of the U.S. CPSC.

Deaths and Injuries from Unsafe Products

Unintentional injury is the leading cause of death among persons under 35 years old and is the fifth leading cause of death in the nation—injuries kill more children than any disease. Injuries account for one out of six hospital days in this country. There are an average 21,100 deaths and 29.4 million injuries each year related to consumer products under CPSC jurisdiction. The deaths, injuries, and property damage associated with consumer products cost the nation about $200 billion annually.

Decline in Deaths

There has been a 20% decline in annual deaths and injuries related to consumer products in the last decade. Agency work in the areas of electrocutions, child poisonings, power mowers, and fire safety helps save the nation almost $6 million annually in health care, property damage, and other societal costs. For example, action on child-resistant cigarette lighters may save over $400 million in societal costs and prevent up to 100 deaths annually. CPSC scrutiny of dangerous fireworks prevents about 14,000 injuries each year.

Hazard Assessment and Reduction

Whenever possible, hazard reduction activities are carried out cooperatively with industry. Since its beginning in 1973, the CPSC has developed over 300 voluntary standards while issuing fewer than 50 mandatory rules; a six-to-one ratio of voluntary to mandatory standards. The 1998 program supports both continuing and new hazard-reduction efforts, including a major initiative on fire hazards. Critical enhancements to the program include a major initiative to address the nation's high fire death-and-injury rate (one of the highest among industrialized nations); an update of the agency's child anthropometric measurements; and funding for significant but unpredictable testing and evaluation support of hazard reduction work.

Reducing Hazards

Recent initiatives of the U.S. CPSC have included the following: window falls, multiuse helmets, movable soccer goals, baseball safety equipment, nighttime bicycle safety, swimming and spa pools, and indoor playground equipment; warnings against the suffocation hazard from soft bedding and efforts to promote back and side-sleeping positions for infants; and outlining cost-effective ways to correct dangerous wiring conditions in older homes. Companies that do not work closely with the CPSC or comply with federal regulations risk money and bad public relations. The CPSC utilizes the media, litigation, recalls, voluntary standards, and other tools to safeguard consumer products. The CPSC relies a great deal upon vigilant consumers who report unsafe products. Additionally, the product-safety community works closely together to provide consumers with the confidence they demand in the products they purchase. Despite this power, the agency does not approach industry with an adversarial attitude. Instead, it has developed a philosophy that works with industry instead of against it, a preference for voluntary actions instead of mandatory regulations, and a belief that programs to educate and inform the consumer are necessary to ensure the public's safety.

The commission works with industry whenever it can help it meet product-safety standards, import requirements, and government regulations. Industry must operate within this framework in order to succeed in the global marketplace. Business should be proactive and responsible; products should be scrutinized before they hit the market. Activities must be defined as the pursuit for quality. Board rooms, rather than technical labs, have to make product-safety decisions. Industry must take initiatives to build safety into its products and must work with CPSC.

Safety Sells

Through partnerships with the CPSC, the government, and nonprofit organizations, industry is building safety and quality into its products. The CPSC helped coordinate the Safety Sells Conference, which recognized progressive corporations such as Hasbro, Whirlpool, and Rollerblade, who place an extraordinary emphasis on safety. These companies learned that quality and safety are good for business. Safety is necessitated by increasingly vigilant consumers, the pressure of organized consumer associations, the fiscal damage produced by litigation, and the high standards and strong regulations enforced by government agencies.

INTERNATIONAL PRODUCT-SAFETY ISSUES

I was impressed in each country I visited by the dynamic changes in the economy, and the change in attitudes of customers, government, standards, organizations, and business leaders, as they related to standards, safety, and other consumer issues. As economies change, the concern for higher-quality products changes.

Today, we see nations around the world very concerned that their families be provided with a safer environment that will allow their families to work and live in surroundings both in the workplace and home that will not cause any unreasonable harm or injury to them. As we are now part of a global marketplace and economy, it is very clear that we must reduce the road blocks and obstacles that prevent us from exporting and importing our products from one country to another.

Report to America

I have talked to many Asian business and government leaders in symposiums, trade missions, and one-on-one meetings. At the request of the U.S. Association of Southeast Asian Nations (ASEAN) Business Council for issues with the ASEAN, I presented symposiums in Malaysia, Thailand, Singapore, and the Philippines; and participated in the U.S. Commerce Department's first delegation on standards to the ASEAN countries. We met with government and business leaders from the aforementioned nations, as well as Indonesia and Vietnam. Asian business and government leaders, and consumer safety issues, are all of importance to the American and to the international marketplace. I asked each of the participants at the International Product Safety Symposiums, which I conducted, a series of questions that were then summarized in a "Report to America." The questions included:

1. Harmonization of standards, will they work?
2. What are the road blocks and obstacles to exporting and importing?
3. Does safety sell?
4. What are the new trends in marketing?
5. How can we be in closest contact with the consumer?
6. How can the U.S. product safety agencies be more helpful to the developing countries?

Harmonization of Standards

Although the harmonization of standards was seen to be very important and would have excellent benefits, it was pointed out that this will take time and be too expensive for some smaller companies. However, it was clear that some smaller requirements of mutual recognition of standards could be very important.

Road Blocks to Exporting and Importing

Road blocks and obstacles to exporting and importing was a major problem. Many concerns focused on lack of knowledge about specific regulations for specific countries; poor communications on safety initiatives; a lack of recognition of test reports from countries; problems with dumping of inferior, unsafe products; high import duties on some products; language barriers; and lack of consistency in manufacturing. Others were concerned about no metric system in the United States, which makes it very difficult to have standardization of measurements in international trade.

Does Safety Sell?

Yes, many of the attendees felt that safety would sell; if it was affordable, if there was proper labeling and consumer education, and if there was good product standards certification. Others felt that safety would sell to markets with a sophisticated technology, where a product can be designed with safety considerations in mind, and the consumers could then be educated on how to use the products. Some were concerned about the cost of building safety into the product. Information dissemination was seen to be the key. Almost every participant felt that government should work closely with industry to maintain safety claims and compliance.

New Trends in Marketing

New trends in marketing consumer products included building good value into products, having products be user friendly, and adherence to international standards. Merchandise needs to be geared towards safety aside from just quality and cost. It needs to offer better quality at a reasonable price with added value, and attractive packaging. Others felt that consumers are looking for higher value and recognized brands, and they want health and environmental claims placed on the products. It was clear that these leaders wanted a balance between safety, quality, and cost.

Closer Contact with the Consumer

As these developing economies grow, it is clear that they are most interested in having closer contact with their consumers. Suggestions included information campaigns; using media more effectively; having closer contact with the importers; utilizing communication technology that will make more data accessible; and for industry and government leaders to support consumerism. Consumers need to be educated about their product purchases. This can be accomplished through questionnaires, after-sales services, telephone and fax numbers on all products, and the creation of a consumer service division within the company and government agency. A toll-free "help line" could be initiated to assist the consumer along with regular seminars on specific consumers issues. Controlled items in some countries should have safety elements built into them. Market research and benchmarking with others was also suggested as being of help to the consumers and the industry.

How the United States Can Be Helpful

The United States, a country that has been focused on safety concerns for well over twenty-five years can be very helpful to the emerging economies. Leaders at these symposiums wanted updated consumer information and comprehensive seminars and training on product-safety issues. It was very clear that the countries did not want importers to dump banned items from one country to another. Ways need to be devised to disseminate new trends in product standards and technologies. The Internet and news media outlets were seen to be a source for assistance. ISO and other international standards organizations were seen to be very helpful. Manufacturers need to be updated on a regular basis. They need consistent information on emerging trends and issues that have been a major concern in the United States, and that will eventually have a major impact on their customers. This could be done through more involvement with trade associations and other media outlets. Injury and accident data was very important and discussions took place on the most-effective way to collect this data.

The Need to Promote International Harmonization of Standards

One of the most important international developments is the harmonization of product-safety standards. The challenge to industry and the product-safety professional is to formulate uniform standards for products so that all people, regardless of country, ethnicity, or race will be assured of their safety. Countries can work together bilaterally or multilaterally. Until product-safety standard harmonization occurs, industry will have to be knowledgeable about the diverse requirements, regulations, and standards in every country with which it does business. This present state of affairs is inefficient and costly. The United States should take the opportunity to assist nations in adopting safety standards, policies, and regulations that are consistent around the world.

Many nations are aware of the need to harmonize their standards and regulations, in order to enhance the competitiveness of their products, gain new markets, avoid regulatory hassles, and increase market share for their products. This awareness has manifested itself in the negotiations for trade agreements such as

Pascal Krieger for ISO Bulletin, reprinted with permission from ISO, International Organization for Standardization, Geneva, Switzerland.

the European Union, the North American Free Trade Agreement (NAFTA) and the World Trade Organization (WTO). International harmonization of standards enhances the exchange of safe, quality products across national borders and reduces injuries associated with such products.

Consumer Policy Committee of the ISO

COPOLCO is a significant organization with a great deal of influence in the issues of consumer protection and product safety. COPOLCO is a very effective voice and organization within ISO. COPOLCO has been developing consumer-oriented initiatives over the last decade. More important initiatives include environmental labeling, product-safety issues, and new initiatives to address service industry quality and standards that will include hotels, financial institutions, and others. The consumer voice within ISO has improved significantly as COPOLCO is now one of the four standing committees that reports directly to the General Assembly.

Priority Items for COPOLCO:

1. Environmental issues including air and water quality
2. Small electronic and electric domestic appliances; motor-operated tools; and mechanical home-use devices
3. Services
    a. Tourism services and hotel classification
    b. Building services
    c. Protection of personal data
4. Child-related products
    a. Child-resistant devices, toys, child-care articles, and playground equipment
5. Smart cards (electronically encoded personal-data cards)
6. Equipment for persons with disabilities
7. Energy performance
    a. Environmental management systems
    b. Environmental and energy labeling
    c. Large, powered domestic appliances
8. Contraceptive devices
9. Protective clothing and equipment
10. Bicycles and sports equipment
11. Public information symbols

Consumer Communiqué from ISO

In addition, the semiannual *Consumer Communiqué* from ISO's Central Secretariat (ISO.CS) publishes recent activities of consumer- and standards-related interests that are being developed at the national and international level. This is an excellent, country-by-country resource of recent and ongoing consumer standards and standardization activities. This communiqué is available from ANSI's CIC or directly from ISO.CS in Geneva, Switzerland.

One example is from a report about the United States. The ANSI board of directors approved a new strategic plan for the CIC to guide ANSI's priorities and activities in order to ensure consumer interests are included in the standards-development process. The mission of the ANSI CIC is to ensure consumer interests are included and protected in the United States and international voluntary-consensus standards processes to help improve the quality of life. The CIC mission statement reflects the mission statement of ANSI to enhance the global competitiveness of U.S. business and to help improve the quality of life through the voluntary standards system. Among the major goals of the ANSI CIC are greater participation of influential U.S. leaders, who represent the consumer, in the voluntary standards process; reestablish American leadership for consumer interests in the international standardization system; expand; and to increase the participation of U.S. consumer-product companies in CIC and ANSI.

Product-safety Policy and Concerns

At a recent symposium of the Organization for Economic Cooperation and Development (OECD, Paris) on "Consumer Product Standards and Conformity Assessment," the European Association for the Coordination of Consumer Representation in Standardization (ANEC, Belgium), said that product-safety policy has been greatly influenced in recent years by a growing number of both regional and global initiatives aimed principally at promoting the greater liberalization of international trade that has affected product-safety policy.

A number of concerns, however, have been expressed by consumer organizations. Consumers have been concerned over the perceived dangers surrounding harmonization. The fear is that the process will lead to harmonization at the level of the lowest common denominator. Experience in Europe has shown, however, that many new European standards have greatly advanced consumer safety in certain areas. The dumping of dangerous products onto markets, where perhaps enforcement and control is less well developed, is another major issue. This happens not only in developing countries, but also in the central and eastern European countries, where demand for western-style goods has increased dramatically, and even within the European Union. Counterfeiting of products such as automobile spare parts and medicines also poses a serious problem.

Consumers are becoming more actively involved in the work of the standards bodies. Previously, the work of the standards bodies has been dominated by industrial interests, but the need for direct participation of all the economic and social partners has been acknowledged at the highest political levels. COPOLCO focused on this dialogue at the international level by holding a joint workshop with ISO's conformity assessment policy committee (CASCO). The special symposium focused on "Consumer Representation in the International Standard Process."

The increasing globalization of trade has had an enormous influence on the development of national, and increasingly regional and international, product-safety regulations. Products can be made at the other side of the world, assembled in another country, and distributed across the rest of the world. As a result, the identity of a manufacturer may no longer be apparent to the consumer. A great deal remains to be done in the area of consumer-product safety. Consumers are better informed and better organized, and can judge and choose the products they want.

The Toy Manufacturers of America notes that how consumer safety standards and conformity assessment are implemented will have an important impact upon the toy industry. In the United States, consumers purchase 2 billion toys

Pascal Krieger for ISO Bulletin, reprinted with permission from ISO, International Organization for Standardization, Geneva, Switzerland.

each year. Only 20% of toy purchases are products made in the United States. By contrast U.S. companies design, invent, engineer, and create the marketing programs for more than 60% all toys consumed in the world. Major U.S. toy companies now sell more products outside the United States than they do inside. There is a world demand for safe, well-made products sold at fair prices.

Take the example of a teddy bear made in the Peoples Republic of China. All of the components, with the exception of thread, come from other countries. The eyes were molded in Japan and fixed ultrasonically by a machine in Korea, the polyester stuffing material came from either Germany or the United States, and the pile fabric was produced in South Korea. All these components were put together in China. The bear was created in the design rooms of a midsized U.S. manufacturer, whose engineers produced the manufacturing and safety specifications for the company's customers in Canada, Mexico, Brazil, Europe, Japan, and the United States.

Labels and hangtags on the toy appear in English, Spanish, and French, and products headed for Brazil will require certification from a recognized American or Brazilian laboratory. For the shipments headed toward Japan, a required test mark was placed on the tags and labels to indicate compliance with Japanese toy-safety regulations. Others headed for an Italian port are distributed throughout the European Union. These must have the requisite CE mark indicating compliance with the European toy-safety directive and standards. Global standards and mutual recognition of conformity assessment would go a long way toward allowing manufacturers as well as consumers to meet their needs.

Removal of Dangerous Products

In order to get unreasonably dangerous products off store shelves and out of homes, the CPSC has instituted major recalls for lead-containing crayons; collapsing tubular-metal bunk beds; flammable skirts, scarves, and fleece garments; and hazardous fireworks. The CPSC has addressed the strangulation risks for children presented by window cords and strings on children's outerwear. These two efforts alone will save 35 children's lives over a five-year period. The CPSC has completed over 400 individual cooperative recalls and corrective actions in 1995.

Also in 1995, CPSC and industry generated 428 recalls, without litigation, involving over 30-million product units. Recalls involving just cribs, cradle swings, and playpens saved an estimated 38 children's lives. The U.S. Customs Service, at the request of the CPSC, prevented an additional 822,000 toy-product units that violated safety standards from entering the country.

Case Example of Noncompliance
U.S. CPSC and Department of Justice versus Big Save International Corporation

A lawsuit was filed by the CPSC and the Department of Justice against Big Save International Corporation in Los Angeles, seeking civil penalties and injunction. Between 1991 and 1995, Big Save imported more than 70,000 banned or mislabeled products including bicycles, baby walkers, pacifiers, rattles, and crayons. They charged that the firm ignored numerous notices of violations from CPSC between 1990 and 1994. Despite the notices, which included warnings of penalties, the firm continued to import products that violated the Federal Hazardous Substances Act and the CPSC's safety regulations. The CPSC said ''. . . that repeated safety violations for the children's products in this case are inexcusable and no firm will be permitted to profit by cutting corners on the safety of our most precious and most vulnerable consumers.'' The banned products included toys, pacifiers, and rattles that separated into small parts when tested and posed a choking hazard to young children. Also, bicycles lacking the required foot brakes and accompanying safety and maintenance instructions. The U.S. Customs Service either stopped or recalled all of the products involved.

SUMMARY

How Can We Be More Responsive to the International Marketplace?

1. We are part of the international marketplace with many new participants. Let us understand, relate, and form partnerships.
2. We need to know the international customer. How does the consumer in Eastern Europe differ from the consumer in Asia and North America? Will our products withstand the close scrutiny of the international marketplace? Ask yourself if you would buy the product before you ship it overseas.
3. We must set a benchmark for quality and safety standards.
4. We must set a goal to take our product-safety processes and efforts "up a notch" in order to achieve excellence in the international marketplace.
5. Business must know the customer, know the trends, and address new concepts and ideas in order to become cutting-edge companies.
6. Industry and standards organizations should consider setting up a consumer advisory committee to the board of directors or executive board. This step is an effective way to learn first hand what the experts—the consumer—thinks and what trends are important.

INTERNATIONAL INSTITUTE FOR CONSUMER SAFETY

A Proposal

I believe there is a need for an International Institute for Consumer Safety that will focus its efforts in the Pacific Area. This institute could become an effective agent for assisting nations to meet and develop product-safety standards for the international marketplace and create consciousness among government, industry, and consumers of the need and importance of safe, quality products. The institute would conduct research and collect data related to unintentional injuries and deaths from unsafe products. The institute would also develop major educational initiatives in cooperation with government, industry, standards officials, consumers, and health officials.

Mission Statement and Vision

1. To promote harmonization of product-safety standards to increase international trade and to enable industry to produce safe products for the global marketplace.
2. To prevent injury and death from unsafe products by educating consumers and working for a world where consumers have to worry less about product safety.
3. To coordinate product-safety initiatives as well as to utilize information and research to assist governments in protecting their citizens and to assist business in developing safe, quality products.

I envision this institute to contribute significantly to the promotion of consumer safety in the international marketplace. To accomplish its goals, it would work closely with business, governments, standards, organizations, and consumers. First, the institute will assist business to produce products that will meet the needs and the demands of the Asian and American consumer. Harmonizing Asian product standards with global quality standards will be a primary objective of the institute. Second, the institute will work with Asian governments to protect their citizenry and to implement effective quality regulations. Third, the institute would work closely with standards organizations to develop standards that meet the needs of government and industry and will satisfy the consumer. It would also assist in the development of harmonized standards, so that products can be placed into the international marketplace more easily. Fourth, for consumers the institute will strive to create a product-safety consciousness to protect the consumer from product deficiencies. Because every consumer deserves the best possible protection, the institute's ability to be proactive is crucial. Media initiatives, symposiums, and consumer-education projects will be utilized to create a progressive consumer-safety movement.

19

Reengineering Standards for the Process Industries: Process Industry Practices

Dr. C. Ronald Simpkins retired in 1996 after 32 years service with DuPont. During his career he served in technical and management roles in engineering, research, plant technical, manufacturing, computer systems, marketing, and business management. In his last assignment, in DuPont Engineering, he was responsible for the administration of the DuPont Corporate Engineering Standards, and defined and led the effort to convert the DuPont Corporate Engineering Standards to full electronic publishing and global server-based distribution. Ron Simpkins conceived, organized, and served as the first chairman of the Process Industry Practices (PIP) Initiative. He served as DuPont's representative on the American National Standards Institute (ANSI) Member Company Council Executive Committee and as a member of the ANSI Board of Directors. Ron is now consulting in the area of electronic creation, publishing, distribution of standards, and standards organizational development, and can be reached at www.simpkicr@dca.net

INTRODUCTION

In early 1993, an organization called the PIP Initiative was formed with 15 companies. Since then it has grown to be a major factor in the process industry. PIP is a joint initiative of 31 major companies, 23 process-industry owners, and 10 contractors who serve the industry. Table 19.1 shows the list of current PIP members. This chapter describes who and what PIP is and why it exists. It will also explain what problems PIP is solving, the benefits and value of this

Table 19.1   Process Industry Practices: Member Companies

3M Company	FMC Corporation
AlliedSignal Inc.	Foster Wheeler USA Corporation
Aramco Services Company	Henkel Corporation
Arco Chemical Company	Huntsman Corporation
BE&K, Inc.	Kellogg Brown and Root
B P Amoco Corporation	Kvaerner Process
Bechtel Corporation	Millennium Inorganic Chemicals
Celanese Inc.	Monsanto Company
Chevron Corporation	Parsons Process Group Inc.
CITGO Corporation	Phillips Petroleum Company
Degussa-Hüls	Rohm and Haas Company
The Dow Chemical Company	S & B Engineers and Constructors, Ltd.
E.I. DuPont de Nemours & Co., Inc.	Shell Oil Company
Eastman Chemical Company	Solutia, Inc.
Elf Atochem North America, Inc.	Texaco Inc.
Fluor Daniel, Inc.	Union Carbide Company

effort, and some of the innovative operational features of the organization. If the reader makes it through this rather weighty list of topics, there is also the author's perspective on what it's like to conceive, organize, and start such an organization.

THE PROCESS INDUSTRIES

Who Are They, What Standards Do They Use, and Why?

Most of us have at least a conception of manufacturing processes in the piece-parts or hard-goods industry. Products such as automobiles, electronic devices, and furniture are assembled as individual units, usually from some level of sub-units or components, and each unit can be uniquely identified and characterized. The products and manufacturing processes of the process industries are quite different, and the industry is not familiar to most of us. Therefore, a generic description of the process industry and the use of standards in that industry is provided.

Standards in the piece-parts and interconnect industries (auto, computers, communications, etc.) typically define product characteristics, such as performance requirements and interconnection specifications. They typically address definition of properties such as size, shape, and electrical characteristics.

Standards used in the process industries are generally not product-related standards. (There are several significant exceptions such as fuels and lubricants.) Standards used in the process industries are primarily engineering solutions that define how to design, construct, and maintain manufacturing facilities. Such standards are used repetitively and have consequentially been refined and optimized. To continue the comparison, the piece-parts and interconnect industries have some process standards, such as those that define the construction of an automotive painting enclosure. In general, the ratio is in order of 90% product versus 10% process standards for the piece-parts industries compared to 10% product versus 90% process for the process industries.

TYPICAL PROCESSES

Typically, raw materials, naturally occurring substances such as natural gas, crude oil, ore, air, and water are purified and reacted to form intermediate products. The intermediate products may be materials such as ethylene, hydrogen, ammonia, or more complex materials such as adiponitrile and adipic acid that are building blocks for more complex materials (such as nylon). The intermediate products may be further reacted on-site, or shipped to another processing location to be converted into final products like polyethylene, nylon, fluoropolymers, elastomers, pigments, or gasoline.

While the raw materials and intermediates may be liquids, gases, or solids at normal temperatures and pressures, the ''products'' of the process industries are typically liquids or solids under normal conditions. From these products, we mold interior automotive parts, make textiles, pigment paints, and fabricate many of the items of everyday life. An interesting characteristic of many of these ''products'' is that until they are ''processed'' into the final form the user sees, the materials are characterized by ''bulk'' properties such as density, color, hardness, etc. In general, individual quantities or volumes are not segregated and characterized by their properties or performance, except as being representative of bulk properties.

INDUSTRY CHARACTERISTICS

Generally, process-industry companies do not make consumer products directly, but rather serve as suppliers to other industries. Again, an important exception is fuels and lubricants. The industry is quite capital intensive when compared to most piece-parts manufacturing companies. For example, it is not unusual for a chemical or petrochemical plant costing $400 million to produce annual revenue of $400 million with a direct operating staff of only 40 people, and a support staff for maintenance and shipping of another 40 to 60 people.

In general in this industry, manufacturers differentiate their products on physical and chemical properties, rather than whether the products meet one or more consensus standards. In fact, most manufacturers have steadfastly worked against the creation of industry product standards because they use the physical and chemical characteristics of their products to differentiate them in the marketplace. It should be noted that there are beginning to be some notable exceptions to this position (e.g., standard nylon for molded-interior automotive parts).

The manufacturing facilities, in other words, the plants, are very "nonstandard." Processing technology continues to evolve and each time a new plant is built, the newest (and most economical) processing technology is incorporated. The individual pieces of processing equipment are uniquely tailored for the individual product and process. The design of this equipment, along with overall process design and operating conditions, are considered critical in any company's competitive advantage.

The processing equipment can be divided into four groups.

1. Reactors. Vessels in which the chemical reactions are carried out. These are very process dependent, may operate at extreme elevated temperatures and pressures, and can be constructed of very expensive materials such as tantalum and special stainless-steel alloys.
2. Separation equipment such as distillation columns and heat exchangers.
3. Equipment to move material such as pumps and compressors.
4. Storage facilities, typically tanks.

Very little of this processing equipment is "standardized," that is, used repetitively enough such that it is selected from a catalog. Reactors are very process dependent. However, even in the case of reactors there are common features such as the size, location, and fittings used in constructing manways that could be standardized with significant cost savings.

Probably the most standard types of equipment are pumps and motors. Defining the details for ordering a pump usually requires filling out at least a full-page "data sheet." The specifics of pumps, including how they are mounted, if standardized could also yield large savings. On examination, it turns out there are many opportunities for the industry to profit from standardizing these design features without compromising competitive advantage. The savings result from the equipment vendors not having to receive and interrupt the various unnecessary details; simplification of manufacturing; and inventory reduction.

Most owner companies, especially the larger ones, have "strategic alliances" with manufacturers of equipment such as pumps, heat exchangers, and distributed control systems. These strategic alliance partners told us that there were significant savings if the industry could standardize on versions of their equipment rather than have a version for each company. They also told us that most of the increased costs (or additional cost differences) they saw were based on preferences rather than real performance differences.

Most owner companies have their own internal set of engineering or corporate standards. *Engineering standards* for this industry means the set of rules used to design, specify, install, startup, and maintain manufacturing facilities. One of the main factors causing each company to have their own standards is that 10 to 15 years ago most larger manufacturing companies did their engineering design, specification, procurement, and construction management internally with their own engineering departments. Now almost all manufacturers use full-service contractors to provide these services, with only minimal oversight by the owners.

The contractors usually have their own set of standards, often a collection of the standards of the owners for whom they have worked. The workload of the contractors, as a group, is very cyclical and there is a large turnover among most contractors, with engineers and support people moving from contractor to contractor. Hence, whereas engineering standards of individual owners were carefully guarded and held secret, now several contractor firms will work with any given owner's standards. Those contractors certainly work to conform to the nondisclosure agreements in place with their customers. However, because most owners deal with multiple contractors, and because of the movement of people between contractors, there is little information really secret in the design, procurement, construction, startup, and maintenance of process-industry plants. Until recently, owner company management in the process industry has not recognized this phenomena.

Note: In one of the initial studies leading to the formation of PIP, several owner's and contractor's sets of engineering standards were examined. It was amusing to find direct copies of several owner standards in several different contractors' sets of standards.

Most process companies are geographically dispersed in the United States and globally. Many have evolved with a high degree of site autonomy with regards to facilities engineering, including project inception and management practices. Not surprisingly, this site autonomy has led to many site-specific engineering standards. Most companies have competent people at their manufacturing facilities, and those people universally argue that "they know what's best for their site." The result is that at many sites across many companies, features are added in equipment design or installation that are based on personal preferences rather than sound, cost-effective engineering.

IN THE BEGINNING: HOW PIP WAS FORMED

In 1992 the author, then a member of DuPont Engineering, was requested to lead a study to determine the strategic direction for DuPont's Process Control Standards work. DuPont had a large, well-defined set of standards that spelled out how instrument and control systems should be designed, specified, installed, and maintained. DuPont was moving away from doing its own engineering, instead it was making alliances with contractors and reducing its internal engineering manpower. There was a wide-spread belief within DuPont that it should not continue to develop, maintain, and use its own standards, but only needed to use industry standards. One of the main concerns was that the DuPont Standards were considered to be ''gold plated,'' that is too conservative, and cause added costs.

A small team of experienced engineers was formed and spoke with contacts in other owner and contractor companies. These informal discussions led to more formal meetings where the Process Control Standards of several companies were compared in detail. Those findings led to the formation of a planning committee made up of representatives from four owners and four contractors. Other studies of a broad range of engineering standards were carried out. The results were surprising. Several key facts were recognized and some important conclusions were reached.

1. Most of the internal standards of these companies were the same technically. Some were presented differently, but contained essentially the same technical content. They were often bulky, contradictory, and sometimes not clear. Extra capital costs were incurred for interpreting them for each project.

2. In most companies, the owner company senior management was saying ''Stop using our internal standards because they're 'gold plated.' '' Just use industry standards. Perceptions were that most of the internal standards were too conservative and cause added costs. In many cases, internal standards represent collections of ''fixes'' piled upon one another. If most companies had an incident several years ago, they made change to insure it didn't happen again. Then they put another, and another fix on top of that.

However, reality was (and is) that there are not adequate industry standards to execute projects in a cost-effective manner. That is, there aren't a sufficient number of ''documented, repetitively used engineering solutions'' to avoid significant reengineering of routine problems. A good indication of the need for standards beyond what the societies offer is that essentially all process companies have large numbers or volumes of internal standards.

3. Most owner companies (or their contractors) also do ''specials''— unique, elegantly designed pieces of engineering that will do the job—but at a higher cost than an equivalent ''off the shelf,'' commercially available solution.

4. It costs money to maintain those internal standards. Most companies feel they are spending too much on their internal standards activity.

5. Most companies have a high degree of site autonomy, and most of those sites have a significant number of site specifics, some of which represent preferences rather than needs.

6. And most importantly, harmonization within the industry is possible and would be very beneficial.

At this point in 1992, the small planning group of eight people tested the water with a number of their acquaintances. Most of the responses were very favorable. Encouraged, an informational meeting was held in late 1992, followed by an organizational meeting in early 1993. The attendees at both meetings represented more than 20 companies. At the January meeting 15 company representatives committed their companies to join this new organization. However, there really was no organization at that moment.

Forming the organization was the next step. Joining with one of the existing standards developing organizations (SDOs) was considered. That option was rejected because of the lack of "fit" with any of the SDOs and the belief that the cycle time to produce a "Practice" could be substantially shortened over the time generally taken by an SDO to produce a standard. Also, none of the SDOs dealt with the wide variety of technical fields where new practices were required. Forming a 501(c)(3) corporation was considered, but rejected because of the amount of energy required to get formal agreement from 15 groups of lawyers.

The solution came from the Construction Industry Institute (CII) who invited us to organize as a separately funded subunit. CII is a research institute of The University of Texas at Austin with 90+ members, both owners and contractors, from a wide range of industries. Most of our PIP members are also CII members. The 15 prospective member companies were signed up, sent an invoice for the first year's dues of $25,000 and the organization effort continued. A strategic plan was developed. The vision, mission, and principles from that strategic plan are included at the end of this chapter. In mid-1993 an office was established in Austin in the same building with the CII offices and a Director and staff were hired.

THE CENTRAL THRUST OF PIP

Figure 19.1 presents a concise view of the objectives of PIP. Currently projects are done with a combination of consensus standards generated by the various societies, internal corporate standards, and those plant or site-specific standards. Most process-industry companies have a large number of internal standards, but many of those internal standards are common across the industry, as represented by the crosshatched areas in Figure 19.2.

CURRENT STATE   DESIRED STATE

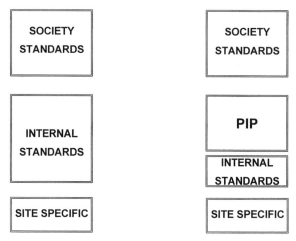

Figure 19.1   A concise view of the Process Industry Practices or PIP objectives.

ORGANIZATION

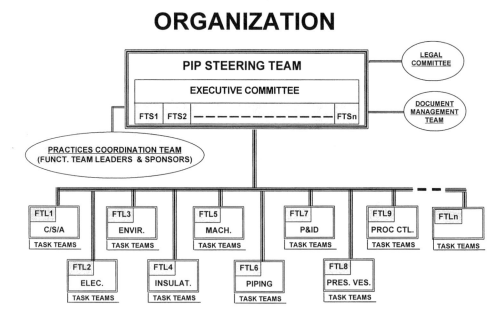

Figure 19.2   Organizational chart of the PIP program.

As this figure also shows, the goal of PIP is to move the common material into our Practices while recognizing that each company will have some fraction of its internal standards that it considers proprietary or special in some way. Society-generated consensus standards will continue to be used where they are available and appropriate, and PIP will work with the SDOs wherever possible. Because of the broad industry acceptance of these practices, and participation by plant people wherever possible, it is expected that over the long term the number of site-specific standards in our companies will be reduced.

The primary task of PIP is to harmonize existing member-company internal standards into a set of industry practices for detailed design, procurement, and construction. Where such internal standards do not exist, and where there is a recognized need, PIP will create the information and issue Practices in those areas. However, PIP will not duplicate work that is being performed adequately by other organizations (the SDOs) but rather PIP will seek to work with and through those groups to identify and converge on a comprehensive set of industry practices. For instance, the PIP Machinery Function Team worked with the corresponding API team to publish a joint standard/practice on machinery installation.

Benefits

Overall the use of the Practices will reduce capital costs by 2% to 5% resulting from the following.

1. Installation details minimize rework and reduce field labor and materials.
2. Standard specification and data sheets reduce procurement costs.
3. Uniform design practices minimize design errors and reduce training costs.
4. Uniform pipeline class identifier and associated pipe codes significantly reduce both design and construction costs.

The Practices are viewed by those in the industry as having improved content and business focus. They are being used in place of internal and site-specific standards and are, in many cases, being adopted as one-to-one replacements for those internal standards.

All of our companies continue to search for ways to reduce costs. By leveraging our company's scarce technical resources some of those people are available for other assignments. Many companies have substantially cut the resources they dedicate to internal standards. The technical people's favorable reaction to their participation in the PIP teams has been surprising. They say that the opportunity to interact with their peers in this setting is extremely valuable and rewarding and have called it the "ultimate technical benchmarking."

Key Organization Features

Figure 19.2 shows a schematic of the PIP organization. It is a very flat organization with only a couple of layers. The governing body is the Steering Team made up of a representative of each of the member companies. The PIP Steering Team is analogous to the Board of Directors of most similar organizations. The Steering Team is led by the Chair and Co-chair, both from member companies, who serve two year terms. Currently, there are nine Function Teams, each led by a Chair. Most of the Function Teams have organized themselves into a number of Task Teams who do the initial harmonization and prepare the first draft to be circulated through the entire Function Team for further editing, if necessary, and approval. Member companies may have a representative on each Function Team and the teams range from about 20 to 30 members.

There is also a Document Management Team with the responsibility of guiding the generation, publishing, and distribution of the Practices. This is a particularly important function as the organization moves into full electronic publishing using the World Wide Web.

Each Function Team has a designated sponsor who is a member of the Steering Team. The role of the sponsor is to promote business value with the team and to provide liaison back to the Steering Team. The Sponsors, along with the Chair, Co-chair, Finance Committee Chair, and the Director serve as an Executive Committee. The Director, who is an employee, is the Chief Operations Officer. The Sponsors manage the office, and a staff of about four full-time people reports to the Chief Operations Officer. The staff is responsible for the final editing and publishing of the Practices as well as administrative support.

Creation and Approval of a Practice

Function Teams are responsible for identifying areas where Practices should be generated. When a team has defined a need, a proposal stating the reason and value, resources required, and schedule for generating the Practice is submitted to the Steering Team. This step not only keeps the Steering Team informed, but recognizes that this team must provide technical people to do the work. When the Steering Team agrees with the proposal, all member companies are notified and requested to submit all internal standards that they want included in the harmonization.

Typically, a Task Team of several people is formed within the appropriate Function Team to harmonize the submitted material and fill out technical voids where needed. At this point, the Task Team members, as representatives of their companies, are encouraged to also seek out any pertinent information from vendors or other outside sources. When the task team has done its work, the draft Practice is submitted to the entire Function Team for further review and approval.

For the issuing of a Practice, the sponsoring Function Team must approve the Practice by a vote of at least 75% of those voting. A further review by the Steering Team is required, since all member companies may not have a representative on every Function Team. The Steering Team must also approve the Practice by 75% of those voting for the Practice to be issued.

The approval process was designed to balance the need for a timely approval versus concerns over reaching consensus. The Practices are primarily intended for member-company use, although they are available for purchase by anyone. Practice(s) are recommendations for voluntary use that may include, but not be limited to, criteria, specifications, procedures, and drawings. They are not put forth as standards. It is expected that user companies will adopt the Practices for use as internal standards, thereby bestowing on the adopted Practices whatever standing that company's internal standards have in use.

Membership

Membership in PIP is open to those companies that have a vested interest in the total quality and cost effectiveness of the process industries, either as a manufacturer qualified for Chemical Manufactuers Association (CMA) membership, or as an engineering design contractor and/or construction contractor serving these owners. The intent is to balance membership with a goal of one contractor for every two owners. Current membership consists of 23 owners and 8 contractors. Member companies agree to share nonproprietary standards, provide members for the Function Teams and have a representative on the Steering Team. Dues are expected to remain at the current level of $25,000 per year.

Results

PIP is very successful by any reasonable criteria. Membership has grown from the original 15 to over 30 member companies. The original 6 Function Teams have expanded to 9 and there are over 300 technical people participating; and these are generally the ''best of the best'' in the industry. Many of the technical people are recognized world leaders in their respective fields.

As an example of PIP productivity, over 220 Practices have been published and another 100 are in progress. The Practices published so far range from 1- to 2-page installation details to a 100+-page treatise on machinery installation. For example, the Civil/Structural/Architectural Team's specifications include ladders, site preparation, concrete, double-contained sewers, and various fabrication details. The Machinery Team developed a general machinery installation guide jointly with API and various equipment-design guides and specifications. The Electrical Team completed specifications for various motors, transformers, and motor control centers. They also have completed a comprehensive set of motor installation details.

The Process Control Team wrote a control systems-documentation specification and developed the first 47 of a complete set of instrument installation details. The Pressure Equipment Team developed a vessel-design guide including specifications, drawing content, and standard details. The Piping Team defined a uniform line-class designator system, and completed the first 43 line-class specifications and 20 of the expected 40+ valve specifications. The uniform line-class designator and associated pipe codes and valve specifications is very significant and exciting. The use of a uniform pipe code system on Process & Instrument Diagrams (P&IDs) and process flow diagrams, even if adopted only by PIP member companies, will produce savings that will more than offset the total cost of PIP.

The published Practices are being used on projects and the initial results of savings are encouraging. A metrics program is in place for the staff to measure use of the Practices on projects. Several member companies have implemented metrics programs internally to measure adoption and use of the Practices, and eventually the economic results.

Availability of the Practices

As stated earlier, membership in PIP is open to any process-industry company and the contractors who serve them. The Practices are supplied electronically to member companies for their unlimited use as part of the benefits of membership. The Practices are available either on paper or electronically for a nominal cost to anyone. To order a catalog, specific Practices, or for information on membership, contact PIP at Process Industry Practices, The University of Texas at Austin, 33208 Red River, Suite 300, Austin, Texas 78705; phone (512) 471-3437; fax (512) 473-2968; and http://www.PIPDocs.org

Some Personal Reflections

The idea behind PIP, the concept of sharing and rationalizing common material so as to eliminate duplication, and the financial incentives for doing so, were never seriously in question. Everyone seemed to believe there would be significant financial returns. What was most in question was whether it was possible to get alignment and real cooperative effort from a group of very competent, but generally opinionated world-class experts, many of whom are appropriately viewed with a sense of reverence within their companies.

It was hard for these people to change their views. At the formation meeting of one of the Function Teams, 12 experts from 12 different companies sat at the table and looked at each other as if to say "Why are the others of you here? I know everything there is to know about xx and yy." (I will not mention the particular technology for obvious reasons.) That particular Function Team expanded to 24 members and became extremely supportive of each other. Even the ones who were originally the most recalcitrant and most likely to pontificate technically have admitted the value of the interaction with their peers. This is typical of the trust and cooperation that has developed within the Function Teams. In this regard, the people of PIP have done an outstanding job.

PIP was and is "an idea whose time has come." As all process-industry companies continue to react to cost pressures by reducing engineering staff, outsourcing functions, and generally running much leaner, the opportunity to reduce internal effort by sharing nonproprietary engineering information is very attractive. The member companies have been very supportive and helpful, and apparently believe strongly in the benefit to be derived from PIP. Most importantly, they have provided technical people to do the work of PIP, producing the Practices. But they also provided leadership and funding, and have enthusiastically spread the word about PIP within the industry. PIP has grown because its success has spread through the industry, not because of a high-powered recruitment or publicity campaign.

I believe PIP has demonstrated that it is business driven, is already producing significant results, and is becoming a recognized force in the industry. There is broad support from the technical community, partly because PIP is seen by many of them as the "ultimate benchmarking experience." I believe PIP is an excellent model for industry cooperation.

ANNEX: PROCESS INDUSTRY PRACTICES—STRATEGIC PLAN

Vision: Voluntary "Recommended Core Practices" for the detailed design, procurement, and construction of process manufacturing facilities are widely available and utilized essentially "as is" among owner, engineering, and construction companies.

Mission: To increase the value of the engineering-procurement-construction process for the U.S. process industry in the global marketplace, and enhance compliance with safety, health, and environmental objectives. This mission is to be accomplished by ensuring the availability of recommended practices for the detailed design, procurement, and construction of process facilities, including the perspective of maintenance and operations.

Definition: The initiative defines *process industry* as those industries, such as chemical and refining, that cause chemicals or hydrocarbon materials and energy streams to interact and transform each other.

Principles

1. Participation is voluntary and open to all companies qualified for CMA membership and the contractor companies serving them.
2. Initial focus is the U.S. process industry.
3. The initiative will support the development of voluntary recommended practices based on a compilation and harmonization of existing member company internal standards, or will develop recommended practices based on new material where harmonization of existing material is not adequate.
   a. Work will not be duplicated that is being performed adequately by other organizations (e.g., ISA, API, ASME). The initiative will seek to work with and through those groups to identify and converge on a comprehensive set of industry practices.
   b. The recommended practices will reference adequate standards that exist and, with the standards owner's permission, those standards will be included in the distribution of the recommended practices.
4. Harmonization or development of voluntary, recommended practices will be done using a streamlined, short-cycle time, defensible (legal, technical) process, which produces practices that are to be used by qualified practitioners (e.g., 80/20 solutions, not recipes).
5. Versions of the U.S. voluntary recommended practices can be used as the basis to meet international requirements.
6. The practices will support implementation of current and anticipated OSHA objectives.
7. Only nonproprietary information volunteered by participating companies or taken from public sources will be considered.
8. A high level of participation, communication, and networking to develop commitment and support for broadly implementing the recommended practices will be expected of participating companies.
9. Participating companies will fund the initiative that will operate on a ''break-even'' basis with lowest reasonable overhead and administrative staff.
10. Recommended practices produced by the initiative will be available in electronic format for purchase by any interested party.
11. The initiative will provide long-term support for updating the practices.

20

Laboratory Accreditation Programs in the United States: The Need for Cooperation

This chapter was contributed by Peter S. Unger, President of the American Association for Laboratory Accreditation (A2LA) headquartered in Frederick, MD. The paper is based on a series of informative presentations about laboratory accreditation including A2LA; and as such, should be recognized that it has an association bias.

INTRODUCTION

Laboratory accreditation is considered in the larger context of conformity assessment. Conformity assessment is the determination of whether a product "conforms" to required standards, specifications, or other descriptors and the attesting to such conformity. The three major third-party tools related to conformity assessment are as follows.

1. *Laboratory Accreditation*: The procedure for recognizing laboratory competence, thereby providing confidence in the accuracy of the data on which conformity assessment decisions are made.
2. *Quality System Registration*: The mechanism that assures the adequacy of a supplier's quality system.
3. *Product Certification*: A source of confidence that a product conforms to specific requirements.

Cooperation in the area of Quality System Registration comes from the American National Standards Institute and American Society for Quality's Registrar Accreditation Board (RAB). Cooperation in the area of Product Certification comes from the ANSI Product Certifiers Accreditation Program. Laboratory accreditation is the one area in conformity assessment where a U.S. national mechanism for cooperation has not yet developed.

LABORATORY ACCREDITATION IN THE UNITED STATES

Unlike most European countries, there are many bodies in the United States that accredit testing laboratories (see Annex). Laboratory accreditation has a long history in the United States, each system generally serving a narrow sector of the economy. The systems are found in the government and the private sector. U.S. laboratory accreditation systems are multiplying without an appreciation for how they are affecting the laboratories, the users of laboratories, or even the exchange of data. Federal and state legislation continues to specify new applications of laboratory accreditation arrangements with little coordination covering the various technical areas affected. Legislation has created separate laboratory-accreditation programs aimed at the medical, aerospace and defense, construction, chemical, and agriculture industries. State legislators are following suit. Some deal with only the test data; others incorporate elements of the broader conformity-assessment process.

Federal and state regulators under general provisions of their enabling legislation are applying various combinations of conformity-assessment features with no overall plan. Government contracting arrangements sometimes call for accredited laboratories.

Likewise, U.S. industry, driven by concerns for safety, performance and the use of the revolutionary ''just in time'' manufacturing principles applies various combinations of the conformity-assessment practices, including laboratory accreditation, to meet their objectives for quality data in decision making and to evaluate and approve testing capabilities of their suppliers. In many instances, these accreditation programs are supported by trade associations or professional societies.

The proliferation of U.S. testing laboratory-accreditation programs has created a patchwork of duplication and redundant requirements. The lack of any overall plan has added to the confusion within the conformity-assessment arena. All these programs deal with the quality of the test data generated by laboratories, whether the laboratories are commercial or part of the manufacturing process. Because requirements are developed by different groups of people, criteria are often different, overlapping and conflicting, and very difficult, if not impossible, to fully implement across the laboratory. With so many interests involved, all with different backgrounds and different levels of understanding of accreditation principles, the situation grows more confusing by the day.

The Government Accounting Office (GAO) has addressed the confusion in laboratory accreditation a number of times, particularly in the environmental area. More recently, a section on conformity assessment was added to the National Technology Transfer and Advancement Act of 1995, PL 104-113, which requires the National Institute of Standards and Technology (NIST):

> to coordinate Federal, State and local technical standards activities and conformity assessment activities, with private sector technical standards activities and conformity assessment activities, with the goal of eliminating unnecessary duplication and complexity in the development and promulgation of conformity assessment requirements and measures.

The origin of this clause of PL 104-113 was based upon recommendations from a report on *Standards, Conformity Assessment, and Trade into the 21st Century* by the National Research Council of the National Academy of Sciences. Their "Recommendation 1" on conformity assessment states:

> Congress should provide the National Institute of Standards and Technology (NIST) with a statutory mandate to implement a government-wide policy of phasing out federally operated conformity assessment activities.

The goal of NIST coordination is the elimination of unnecessary duplication and to promote reliance on the private sector. But the private sector must coordinate and harmonize its activities if it is to meet the challenge. There is much support for establishing a coordination mechanism in order to recognize accreditors and eliminate unnecessary duplication of assessment and accreditation. The Fastener Quality Act mandates that NIST establish a program to accredit accreditors, both foreign and domestic, but it is just being implemented and for fastener testing only. More recently, the establishment of the National Cooperation for Laboratory Accreditation (NACLA) seems to be the solution to this problem.

NACLA is the creation of the Laboratory Accreditation Working Group (LAWG). ANSI, in cooperation with NIST and ACIL (formerly the American Council of Independent Laboratories), started LAWG to find ways to bring laboratories, accreditors, and users of laboratories together. Several laboratory accreditation bodies had already signed a memorandum of understanding in which they agreed to work toward establishing a Mutual Recognition Agreement (MRA). This effort has been merged into NACLA. The model developed by the International Laboratory Accreditation Cooperation (ILAC) could be used as the basis for this agreement. This model is now being implemented in Europe and the Europeans are urging the establishment of MRAs in other parts of the world and eventual bilateral agreements among MRA groups.

The need for a strong private-sector agreement on basic concepts and for a new spirit of cooperation among U.S. accreditors seems apparent. Such an agreement would lead the way for cooperation with government systems in ensuring that data from U.S. accredited laboratories are accepted in the United States and throughout the world. Accreditation systems may soon need to consider greater cooperation to minimize duplication of efforts in their client laboratories. We have already reached this stage in quality system registration. Marketplace forces are causing registrars of quality systems to accept registrations of suppliers from competing organizations because it is inconceivable that one registrar can demand that all suppliers to a manufacturer be registered by that registrar. In order to foster cooperation, an organization of registrars has been established to explore how this recognition among systems will be accomplished.

The members of LAWG believed there is an urgent need to form NACLA, consisting of representatives of any and all U.S. operating laboratory-accreditation systems whether public or private and major users of such systems. NACLA should bring together the community of interests to:

1. Identify common interests and share perceptions;
2. Receive information on national programs and any other provision related to laboratory accreditation;
3. Promote public- and private-sector laboratory accreditation partnerships in the national interest;
4. Assure that laboratory accreditation interests are fully represented in international and other public national forums;
5. Project private-sector solutions in the interest of improved efficiency in laboratory assessment and more universal acceptance of the test data;
6. Advise government and private-sector bodies of the needs in the area of laboratory accreditation;
7. Sponsor related educational activities;
8. Negotiate with national and international accreditation systems and users of test data for acceptance of the results of tests performed by laboratories operated by NACLA members;
9. Consider sponsorship of a means (such as mutual-recognition agreements) to identity laboratory accreditation systems that meet international criteria; and
10. Explore the feasibility of establishing a laboratory accreditation MRA group in the United States that could negotiate with other MRA groups in Europe and Asia so as to establish a formal mechanism to obtain acceptance of test data around the world.

PRIVATE-SECTOR ACCREDITATION OF LABORATORIES— A CASE STUDY

A2LA was formed in 1978 to provide a comprehensive one-stop accreditation program. The idea was to accredit laboratories for their full range of testing capability, not just for a set menu that had to be announced in advance. A2LA is now the largest multidiscipline laboratory-accreditation program in the United States and fourth largest in the world. A2LA operates its program in close harmony with other national programs and has mutual-recognition agreements with several foreign national programs. In addition, A2LA's Environmental Lead (Pb) program is recognized by the Environmental Protection Agency (EPA); its Electromagnetic Compatibility (EMC) program is recognized by the Federal Communications Commission (FCC); and A2LA has a mutual-recognition agreement with the U.S. Naval Sea Systems Command. A2LA is actively pursuing agreements in Europe and the Asia-Pacific region. The existence of A2LA has slowed but not stopped the proliferation of narrow-focus laboratory accreditation programs in the United States.

A2LA: A Membership Society

A2LA is a nonprofit, professional society with individual, institutional, and organizational members. Membership is open to anyone. A2LA's goal is to provide a comprehensive national laboratory accreditation system that establishes widespread recognition of the competence of our accredited laboratories. A2LA also hopes to eliminate the unnecessary multiple assessment of laboratories.

A2LA has been accrediting laboratories since 1980, using the international standard International Organization for Standardization and International Electrotechnical Commission's ISO/IEC Guide 25 "General Requirements for the Competence of Calibration and Testing Laboratories" as its criteria for accreditation. This standard not only requires a quality system and manual in the laboratory but also requires that the laboratory be found competent to perform specific tests and types of tests. In 1990, Guide 25 was revised, considering the content of ISO 9002 (Quality Systems—Model for Quality Assurance in Production, Installation, and Servicing) and is presently undergoing another revision. A2LA's system is designed to meet the requirements of ISO/IEC Guide 58, "Calibration and Testing Laboratory Accreditation Systems—General Requirements for Operation and Recognition."

The association is governed by a 22-person Board of Directors made up of individuals from government, users of laboratories, and those who run the laboratories. Past chairmen have come from government agencies, the auto industry, the coal testing industry, a public power utility, and an insulation manufacturer. A2LA uses two councils for its decision-making process: the Accreditation Council for decisions on accreditation actions and the Criteria Council for decisions on accreditation requirements. Funding comes from membership dues, fees for services, and training courses. The current budget is over $4,000,000.

A2LA now has over 1,150 laboratories accredited, growing at the rate of about 15 per month. Major fields of testing include mechanical, chemical, construction materials, electrical, and environmental testing. A2LA currently accredits nine calibration laboratories and actively participates in the Asia-Pacific Laboratory Accreditation Cooperation (APLAC) proficiency-testing programs for calibration and testing laboratories.

One of the strongest supporters of A2LA has been General Motors. They have a program to accredit supplier laboratories, based on Guide 25. But, they do not accredit second-, third-, and fourth-tier suppliers, nor commercial laboratories that serve the industry. They recognize A2LA accreditation of these organizations, mostly in the mechanical and chemical testing areas, but now also in environmental testing. Similarly, companies like Shell Oil have come to rely on accreditation in the environmental area in lieu of doing their own assessments.

Registration of Laboratories to ISO 9000

With the worldwide acceptance of the ISO 9000 series as the quality system standard for all types of endeavors, including laboratories, some of A2LA's laboratories accredited to Guide 25 are being urged by their customers to obtain registration to ISO 9000 as well. A2LA has been asked to provide the service. Consequently, a new document, entitled "Requirements for the Registration of Laboratory Quality Systems," provides a process for registering laboratories to ISO 9001 or 9002. Two types of recognition can be obtained:

1. Accreditation of competence of a laboratory based on ISO/IEC Guide 25
2. Registration of laboratory quality systems based on ISO 9001 or 9002

To become registered (2), the laboratory must first become accredited (1) because the determination of competence is fundamental to the association's work. The additional items that need to be added to meet ISO 9000 requirements and the requirements of quality system registration as practiced worldwide are included in the laboratory-registration program.

Additional Requirements for Quality System Registration

1. A more complete assessment of the quality system documentation and agreement on its adequacy before the on-site audit. More auditor time is needed—at least one-half day.

2. A preaudit visit by the lead auditor is encouraged for laboratories seeking accreditation and registration.
3. On-site audit according to ISO 9002 (or equivalent 9001) paragraphs.
4. All initial and renewal audits generally done by at least two people, with one member of the team being a certified lead auditor. Accreditation assessment and registration auditing are to be combined wherever possible.
5. An on-site surveillance visit must be conducted at least once per year unless a complete reassessment or audit has taken place. One auditor for the surveillance is the norm; normally at least two for reassessment or audit.

There are two applications: one for accreditation and a second for registration. A2LA has achieved RAB accreditation of its registration programs for laboratories and laboratory-related suppliers.

Registration of Reference Material Suppliers

At the request of the EPA, A2LA developed a program to certify reference materials to meet EPA-A2LA jointly developed product specifications for neat, synthetic, and natural-matrix reference materials. The program requires reference materials suppliers be registered by A2LA to the ANSI/ASQC Q9001 or Q9002 (ISO 9001 or 9002) quality system standards. Each supplier meeting these requirements is formally registered and may advertise the fact of registration. A2LA's RAB accreditation includes the registration of reference material suppliers.

Certification of Reference Materials

Each lot of reference materials for which a supplier seeks certification is accompanied by test data from a laboratory meeting the requirements of ISO/IEC Guide 25, "General Requirements for the Competence of Calibration and Testing Laboratories" and supported by data from referee laboratories meeting those same requirements. Data from each lot are reviewed by an A2LA-retained technical expert to see that they meet the specifications. When they do, the supplier is granted permission to use the A2LA Certification Mark on the relevant lot.

A2LA Training Programs

A2LA offers a number of training courses related to its activities. Currently, these courses include Guide 25 and Accreditation, Documenting Your Quality System, Internal Audit, Measurement Uncertainty, and Calibration Laboratory Practices.

In conjunction with the National Association of Testing Authorities (NATA), Australia, A2LA offers Assessment of Laboratory Quality Systems. The latter course is a five-day lead-auditor training course registered by the Institute for Quality Assurance (IQA in London), which combines ISO 9000 and Guide 25 in its study materials. These courses have been presented nationally and internationally.

CONCLUSION

It is in programs such as these that the private sector serves industry in the United States. Government policy in the United States encourages government use of the private sector wherever practical. A2LA has petitioned to various government agencies including NIST for recognition of its various programs that support regulatory and procurement decisions and trade. But A2LA is also prepared to enter into agreements directly with governments as has been done in the case of Hong Kong (PRC), the U.S. Navy, EPA, and FCC. Being a membership society, all are invited to join A2LA. Members receive the *A2LA News*, the A2LA Directory of Accredited Laboratories, the Annual Report, and discounts on training and other services.

ANNEX: SOME CURRENT LABORATORY ACCREDITATION SYSTEMS AND APPLICATIONS

Federal legislation covers the following areas:

Medical: The original Clinical Laboratory Improvement Act (CLIA 1968) directed the Health Care Financing Administration (HCFA) to create its own laboratory accreditation program and at the same time to accept already existing programs of equal rigor. It worked for social security payments to doctors, but affects less than 10% of the medical laboratories.

Doctors' offices: New legislation puts tough accreditation requirements on all doctors' office laboratories, estimated to be 200,000. HCFA has created very complex regulations to implement this law, taking little advantage of the wealth of international guidance that is available.

Fasteners: The Fastener Quality Act sets up laboratory accreditation as the method to control fastener quality. In this case, NIST must establish its own system to accredit laboratories and it also must set up a procedure for recognizing other laboratory-accreditation systems operating in the private sector.

Lead: The U.S. EPA has established a program for accrediting laboratories for testing lead in paint and other environmental media related to the lead remediation efforts, recognizing A2LA and AIHA as private-sector accreditation programs.

Pesticides: The Department of Agriculture and the Food and Drug Administration (FDA) have been directed to set up a laboratory-accreditation procedure for those laboratories testing for residual pesticides in food. In this case, the Department of Agriculture has been directed by legislation to use private-sector bodies where possible.

Asbestos: Congress directed that all laboratories that test asbestos in relation to asbestos-elimination projects in public schools must be accredited by NIST. This may or may not be extended to other construction sectors in the future.

Federal regulation designed by agencies under the more general provisions of their enabling legislation is resulting in the establishment of more laboratory-accreditation programs. Some examples are:

Agriculture: There are at least six programs operating in the Department of Agriculture with little coordination among them.

Food and Drug Administration (FDA) Toxicology: The FDA originally developed the Good Laboratory Practices (GLP) for use in recognizing the adequacy of long-term animal-feeding studies in the assessment of toxicity of food products. They talk of extending the idea to broader testing areas.

Environmental Protection Agency (EPA): EPA adopted the FDA approach for its toxicity studies and had a key role in promulgating the whole idea to the international arena via the Organization for Economic Cooperation and Development (OECD). EPA has developed accreditation procedures under the National Environmental Laboratory Accreditation Conference composed of government officials at every level to coordinate environmental laboratory accreditation among the states and other agencies.

Federal Communications Commission (FCC): FCC has procedures to minimize radiation interference from consumer products. FCC is accepting accreditations performed by NIST/NVLAP and A2LA.

Department of Housing and Urban Development (HUD): HUD has for years distinguished between laboratory accreditation and product certification and accepts both government and private-sector accreditations. A new program related to lead-based paint is being implemented in cooperation with EPA.

National Institute on Drug Abuse (NIDA): This program for accrediting testing laboratories for drug measurements is being implemented by the agency with little use of laboratory accreditation already being used in government and elsewhere.

Bureau of Customs: A program has been set up to accredit laboratories that test items whose ingredients affect the tariffs charged on imports. In this case, customs decided not to use other available systems and to do the accreditations themselves.

National Voluntary Laboratory Accreditation Program (NVLAP), NIST: This program was designed for use when a related private-sector program will not serve the need. In some cases, such as the asbestos and fastener cases, NIST/NVLAP have been written into the legislation. In other cases, such as insulation and construction materials, the program was established before a private-sector program was available.

Federal contract requirements for the use of laboratory accreditation are being established. Examples include those which follow.

Army Corps of Engineers: The Corps does the assessment. They will accept assessment done by various public and private-sector programs operating in a related area.

Naval Facilities Engineering Command (NVFAC): Relies on NIST/ NVLAP, NIST/CCRL, and A2LA programs.

Military Supply Centers: Some centers have their own extensive programs. Others have a laboratory listing service and encourage their contractors to be accredited by private-sector organizations.

Federal Aviation Administration (FAA): FAA is accepting private-sector accreditations.

State and local agencies implement accreditation programs. Over 50 systems involving most states serve as examples. These systems are primarily in two areas: construction materials and environmental (drinking water).

The private sector has created many separate laboratory-accreditation systems with varying types of requirements. Some examples are:

A2LA: This is a multipurpose system designed to serve a variety of users.

American Industrial Hygiene Association: Administers several accreditation and proficiency testing programs that recognize laboratories performing industrial hygiene and environmental analyses serving users in the public and private sector.

American Petroleum Institute: Administers an accreditation program for tests on various oil products.

Chemical Specialties Manufacturers Association: A trade association program focussed on pesticides, which declined to use available systems in favor of establishing their own.

College of American Pathologists: Originally accepted under CLIA in early 1970s, there are six other programs in this area; 20,000 labs accredited.

Fenestration Rating Council: Include aspects of product certification. Declined to use existing systems and is creating a new, competitive system for testing windows and doors. This effort is funded by the Department of Energy.

General Motors, Chrysler: Used to support just-in-time manufacturing processes, where in order to accept components or subassemblies, the supplier must have their laboratories supply data showing compliance with accreditation criteria. They do their own assessments but are also accepting other private-sector systems.

International Conference of Building Officials (ICBO) Evaluation Services: Used in support of approval of products used in structures covered by building code requirements.

International Safe Transit Association (ISTA): Accredits packaging testing laboratories.

Society of Automotive Engineers (SAE) Performance Review Institute: Sponsors several programs in specific testing areas such as nondestructive testing, mechanical, and chemical to address aerospace and defense industry needs.

21

Test Methodology and Standardization Parables

The Passenger Car Tire Committee of the Society of Automotive Engineers (SAE) developed a Recommended Practice (SAE J918) to provide minimum requirements and "uniform laboratory test procedures for evaluating certain essential characteristics of new tires intended for use on passenger cars." One of the characteristics determined was the ability of the tire to carry safely the load imposed on it by the vehicle, its passengers, and cargo. This chapter on tire testing and other standardization parables by Dr. F. Cecil Brenner is reprinted in part with permission from *ASTM Standardization News*, Vol. 1, No. 9, 1973, copyright American Society for Testing and Materials, 100 Barr Harbor Drive, West Conshohocken Drive, Philadelphia, PA 19428.

TAMPERING WITH A STANDARD

The method of testing involved running the mounted, inflated tire against a (67.23 in. or 1.71 m in diameter) steel wheel (1/300th of a mile or 1/480th of a kilometer in circumference) in an air space controlled to $100° \pm 5°F$ (38°C) (Fig. 21.1). The tire was pressed against the steel wheel with a load it was to be qualified to carry. Thus, if the tire was to be rated to carry a specified load it was to be pressed against the test wheel with the specified load during the first phase of the test. Then the tire was run at a constant 50 mph (80 km per hour) for four hours followed by six hours and 24 hours at 120% and 140% of the specified load. Tires that ran through the procedure without failure—blow out or separation—were qualified to carry the specified load. In practice, the auto companies always selected tires with somewhat greater load-bearing capacity than required.

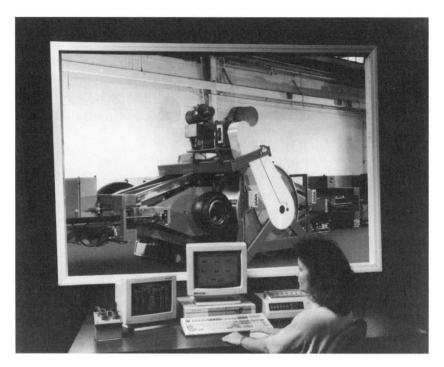

Figure 21.1 Test for tire endurance.

This test procedure was used for a number of years to good effect. Tires meeting the standard gave satisfactory performance with regard to their load-bearing characteristics. Indeed, many not only served satisfactorily during the period of their original tread's use but were capable of being recapped with a new tread and again serving satisfactorily. This empirical evidence shows that the standard served its purpose very well.

In June 1965, the SAE Tire Committee tampered with this procedure with calamitous results. They specified that a tire was to be tested so that the "average radial deflection of the tire" at the specified load "shall be determined on a flat surface. . . ." (The change was described in a footnote to a table specifying the tire test loads!) This change dictated that a tire be loaded on a flat surface to the specified load and have its deflection determined. (The deflection is the difference between the radius of the unloaded tire and the distance from the center of the axle to the contact point on the flat surface of the loaded tire.) Now the tire was pressed against the steel wheel with the load that produced the flat plate deflection. It turns out that the deflection on the wheel was achieved at about 88% of the load on the flat surface. The result of this change was that a tire rated to carry, say 1,000 lbs. in previous years was tested at 880 lbs. after June 1965 but still was rated to carry 1000 lbs.!

To understand the possible motivation for this tampering one must be aware of two facts:

1. Tire size and cost are related—the larger the size the greater the cost.
2. Tire size and load-bearing capability are related—the larger the size the greater the load-bearing capacity.

As a result of the change, many model year 1965 automobiles that were fitted with a tire of a certain size could, for model year 1966 be fitted with a smaller sized tire. For example, a 1965 vehicle requiring say a 6.90-14 (bias-ply tire-size designation) could be fitted with a 6.50-14 in 1966 on a vehicle of the same weight. If the cost difference was an average of $0.50 per tire the cost savings would be have been $2.50 per vehicle (five tires per vehicle in those days). The auto industry savings in a $9 million–sales vehicle year could have been $22.5 million.

What were the consequences of this tampering? Many (new 1966 model) car owners experienced several tire failures at relatively low mileage (less than 10,000 miles; 16,000 km) and complained to Senator Gaylord Nelson, Wisconsin, who earlier had spoken out about the safety of tires. The avalanche of letters received by the senator may have contributed to his activity to assure that there be a mandatory requirement for safety standards for passenger car tires included in the Motor Vehicle and Highway Safety Acts of 1966. In January 1966, the SAE Passenger Car Tire Committee again revised the test procedures by deleting the previously mentioned footnotes.

MULTIPLE TESTS FOR THE SAME CLASS OF PRODUCT FAILURE

This section illustrates what I have called the Bed of Procrustes Syndrome in test-method development, and also explains why there are frequently several tests for the same class of failures. It should be noted that the motivation for developing a test for a manufactured product is to assess its performance. A good test allows the development engineers to assess performance in the laboratory and avoid the expense of making field tests until the problem is thought to be solved. This part is reprinted from an article published in ASTM's *Standardization News*, Vol. 1, No. 9, September 1973, pp. 11–12.

The Bed of Procrustes' Syndrome

There is an analogy between simulation test developers and the fable of the ancient Greek bandit, Procrustes. Procrustes' kidnap victims were forced to lie in a bed; if they were too long, he cut off their legs, and if they were too short, he stretched them until they fit the bed (Fig. 21.2). The test-method developer frequently is cast in the role of Procrustes. Here the test method is analogous to Procrustes' victim and the failing or deficient product is analogous to the bed. In other words, the test method must be made to fit the bed. That is, it must reveal the product failure or deficiency or it will not be a satisfactory test.

How this could come about and what the consequences are, are best illustrated by the following fable. Consider a tire system (cord-adhesive-rubber, Fig. 21.3) and its resistance to fatigue failure. Fatigue is a name given to a class of tire failures resulting from continued flexing in use. There are many possible causes for tire fatigue, among which are cord failure, rubber failure, adhesive failure at the cord surface, and adhesive failure at the rubber surface. Now, as

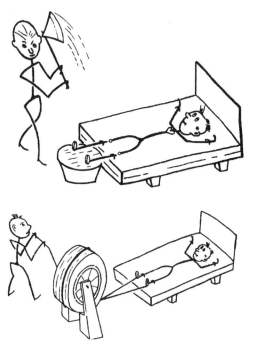

Figure 21.2   Analogy of the Bed of Procrustes Syndrome to tire test method development.

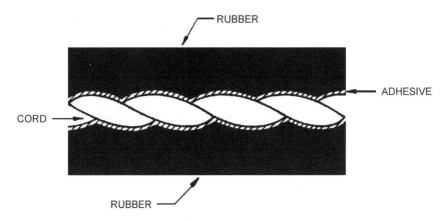

Figure 21.3   Tire cord embedded in a rubber casing with adhesive bonding.

part of our story, suppose that at some time the tire system developed by a particular company primarily failed at the rubber-adhesive interface. Also, assume that a device that we can call Fatigue Tester A was successfully developed and correlates with the system's performance in road tests and in use. Although it was not understood by the inventors, this test was successful because it tended to concentrate stress at the rubber-adhesive surface.

In the course of the company's research activities, a new cord-adhesive-rubber system was developed and the new system failed in the rubber. The data generated by Fatigue Tester A no longer correlated with the road test data and Fatigue Tester B was invented. The new device flexed the sample in a way that caused the stress to be concentrated in the rubber, and its test data correlated with road test and use data for the new system. Across the street, a competing company's cord-adhesive system failed because it was temperature sensitive. Fatigue Testers A and B failed to produce correlatable data for this system because the samples did not develop temperatures as high as those existing in tires in use. So, a new tester (C) was invented to meet this need.

In this example, the three test methods have been forced to fit different beds of Procrustes and each has a real value as a test method if we know what it is trying to tell us. In situations where complex systems such as tires are subjected to fatigue, abrasion, or impact, it is not likely that one test will meet all the needs for complete evaluation. Indeed, it seems inevitable that a variety of tests must be developed for evaluating complex or composite materials that can fail for a number of reasons from the same kind of stimulus. Each test, although valid, can only be used intelligently if the mechanism of failure for the specific case is understood.

Our parable suggests that a simple laboratory simulation of the destructive action is not a satisfactory way of uniquely defining a system failure. Complex systems fail in a number of different ways and each simple laboratory test method is likely to cause the system to fail in a specific way. The existence of a variety of tire fatigue testers that do not correlate with each other plausibly supports this contention.

Evaluation of the resistance of a tire system to flex fatigue requires an understanding of the mechanisms of failure induced by each different test method and an ability to interpret all the results in terms of use experience. The attempt to translate the numerical value obtained by a given laboratory test method into a prediction of overall performance is doomed to failure. The systems are too complex and the number of mechanisms for failure too numerous to permit this kind of solution.

It is contended here that many simulation tests can be used effectively only if the response of the system in the tests and in actual use is understood in detail. Besides leading to improved ability to interpret the results of simulation tests, the research worker will be led more rapidly to the solution of the problem facing him.

LACK OF CORRELATION BETWEEN TESTS

Frequently, the test results from two or more tests for the same class of performance properties fail to be correlated. In the previous section we discussed the development of several "fatigue" testers that stress samples in different ways and, of course, no correlation exists. Sometimes two methods testing for the "same" property do not correlate but, nevertheless, both are useful in different aspects of the general problem.

We may illustrate the point by considering methods used to evaluate the wrinkling performance of fabrics. In one method, fabric strips (1 cm $\times$ 4 cm) are creased and compressed under controlled conditions of time and load. Two specimens are tested; one deformed at right angles to one set of yarns and the other at right angles to the other set of yarns. Each strip is then clamped on a shelf so that the half below the crease hangs freely. The shelf and clamped leg are periodically moved as the recovery takes place so that the hanging leg is almost always vertical. After five minutes the Crease Recovery Angle is determined.

In the second test, a fabric sample is formed in a cylinder and enclosed in a latex cylinder sealed at one end. The fabric and diaphragm are connected to a fitting attached to a pump. The air is alternately evacuated and returned to the diaphragm; the system collapses and expands, wrinkling the fabric. Finally, the fabric sample is removed and laid on a flat surface for a specified period and subjectively evaluated for its wrinkling performance.

The results from the two tests are not correlated. The methods differ in the manner in which the deforming stress is administered. In the Crease Recovery Angle method, the fabric is forcibly creased across each set of yarns regardless of the nature of those yarns. In the diaphragm method, the fabric accepts the stress in accordance with its mechanical properties; the stiffer the fabric the more resistant it is to wrinkling. Since wrinkling performance is dependent on wrinkle resistance and recovery from wrinkling, the diaphragm method is a better method of evaluating wrinkling performance than the Crease Recovery Angle method. However, this latter method provides useful information when comparing the effect of various fabric treatments designed to improve wrinkling performance of fabric.

CORRELATION BETWEEN LABORATORIES

The Crease Recovery Angle Test, was widely used to good affect in many laboratories. An extensive study among eleven laboratories showed that each could reproduce with good precision their results among their technicians (intralaboratory correlation) but with no agreement among laboratories (interlaboratory correlation). A study was undertaken in which the relative humidity, the creasing load, time under load, and other factors were rigidly controlled to determine if interlaboratory variation could be improved. No improvement was found. In a second experiment, additional variables were controlled to no affect. These experiments have been reported by this author ("Test Method Standardization—A Dilemma," *Clothing and Textile Research Journal*, Vol. 3, No. 1 1984–1985, pp. 41–45).

Experiments of the type reported here and their results are not unique. It is likely that many test methods cannot be made to produce the same numerical result on the same material in all laboratories. We have reasoned that this may be true because of the difficulty of describing the procedures that all operators perform the same way. Even when samples are handled exclusively by mechanical instruments that are free of human factors, frequently we will find consistent differences between results on different machines that are almost identical. The differences are likely to be greater the more complex the machine simply because there are more ways in which the machines can differ.

There may be other reasons that make it difficult or impossible to develop procedures that produce uniform results everywhere. For whatever reason, it is senseless to pursue the goal after a reasonable and intelligent effort has been made and has failed. I would point out that the experiments reported here required over three years to complete and cost approximately $90,000+ in 1970 dollars (based on estimates of costs to test 2,982 specimens, computer time, travel, and time spent in meetings of committees, writing reports, etc.).

Two possible solutions to the dilemma have been suggested: (1) establish a set of standard materials, enable each laboratory to determine how it relates to the average experience, and then adjust its results to that of the average; and (2) establish a course so that all technicians are trained by a common instructor and are drilled in common practices of handling and manipulation of samples. Other test methods may require other solutions. The point is, we should not hesitate to develop and use such solutions when the more desired solution is too expensive and time consuming to achieve. The American Association of Textile Chemists and Colorists (AATCC) opted for establishing a training course for technicians apparently to good affect.

SCREENING TESTS

A standard test method may be designed to show that a product will exhibit a required property in service if the test result exceeds a certain value. Usually, the method is developed on products known to be successful in use and it works very well on such products. However, in the course of research and development to improve the product new materials are used. The test method may produce puzzling results. For example, a cotton-reinforced rubber drive belt on a machine is found to perform well if its elastic modulus exceeds 10 lbs. of stress per inch elongation. New belts with increased abrasion resistance are desired and tests on a variety of belts are undertaken. Not all belts with elastic modulus in excess of 10 lbs. per inch give satisfactory performance. Perhaps 10% to 20% of samples that meet the required elastic modulus are satisfactory. Incidently, only belts that exceed the required elastic modulus perform acceptably.

Here we have specified a necessary property for success but one that is not sufficient for success. Obviously, another property or properties are controlling. It may be that belts cannot perform satisfactorily if they stretch during use and do not recover to their original length. A complete test protocol will require a test to assess this property.

Thus, a screening test is one that defines a necessary property that in itself is not sufficient for a successful product.

AN AESTHETIC PRODUCT WITH ONLY SUBJECTIVE STANDARDS

Synthetic textile-fiber producers are frustrated by the inability of fabric designers to tell them specifically what fiber properties are needed to produce the fabric they desire. Development of the elastic fiber and its potential for use in foundation garments—brassieres and girdles—provided hope that the designers could supply such information. Certainly, we reasoned, these garment designers would know the mechanical properties they needed to restrain and shape the human figure.

To exploit this possibility, we arranged a visit to a manufacturing plant of one of the principle foundation garment manufacturers to talk with their technical staff (the quality control director). He listened to the explanation of our purpose with some amusement before dashing our hopes. The designers do not know what fiber properties are needed he assured us. Their approach is just the opposite. They feel, pull, and crumple up the various fabrics then let the fabric properties suggest its use. For example, one of the elastic fabrics submitted was one that could be stretched to eight to ten times its length. This suggested to one of the girdle designers a novelty girdle that could be carried in a purse and was called "the postage stamp."

Was there no objective mechanical data generated in the process? Yes. When a garment's design was complete, the quality control department measured some of the mechanical properties—elastic modulus, stiffness, etc.—of the fabrics use by the designer. (A brassiere has perhaps eight different fabrics; a girdle fewer.) Since the designer is responsible for the product and its success, the designer is then given a number of samples of each fabric with slightly different properties and asked to select all those fabrics that would be acceptable.

The mechanical properties of the acceptable fabrics were measured and the extreme values of these provided the upper and lower bounds for acceptance. Experience had shown the quality control director that the designer's view of acceptability was not constant. In order to avoid conflict the designer was asked to examine samples each month and again select those that were acceptable.

It is not possible to take one brassiere and scale its component parts up or down to produce the many sizes required. However, it is possible to scale up and down from one size to cover several similar sizes. It turns out that each designer specializes in either large-, intermediate-, or small-busted women and the other member of the designer team, the model, reflects this.

In the course of our visit, I was prompted to ask our host, "Do you have only one brassiere designer and one girdle designer?" "No," he responded, "we have 26 brassiere designers and eight girdle designers. Where did you get that idea?" "Well, each time you referred to the brassiere designer you said, "He does. . . ," or to the girdle designer, "she does. . . ." Are all the brassiere designers men and the girdle designers women?" He was startled, "I never thought of it but that's true." I leave rationalization of this to the Freudian psychologists.

REFERENCES

1. WV Cropper. Standards and Standardization. Chemtech, September 1979, p 550.

22

Performance Testing and Correlation

In Chapter 2 we pointed out that standards based on performance tests do not stifle innovation or creativity as do standards that require specific designs or materials. In performance tests, the only method possible is one in which real-world performance is simulated in a laboratory or field test. The designer of a performance test must always demonstrate that there is a correlation between the standard test data and real-world data. This chapter by Dr. F. Cecil Brenner presents an orderly set of steps by which a performance test is developed and shows how one determines if there is correlation between data obtained under different conditions. This chapter is reprinted with permission from *ASTM Standardization News*, Vol. 6, No. 1, January 1978, copyright American Society for Testing and Materials, 100 Barr Harbor Drive, West Conshohocken, PA 19428.

> *Webster's Third International Dictionary: Performance test*, n: a test of capacity to achieve a desired result. *Oxford English Dictionary: Test*, 4b: Mechanics, etc.: The action by which the physical properties of substances, materials, machines, etc., are tested, in order to determine their ability to satisfy particular requirements.

The development of a performance test for a physical property follows a pattern, consciously or unconsciously. The pattern has been described by Rothwell (1) and is briefly reviewed here. Rothwell sets forth four hierarchical stages in the development of a performance test. These are the establishment of performance requirements, criteria, evaluation, and specifications. Upon completion of these stages the fifth step, the establishment of performance standards, is possible. Rothwell described these stages as follows:

Performance requirements: The primary level begins with an analysis of performance requirements. These are qualitative statements that describe the problem for which a solution is sought.

Performance criteria: The second level identifies the performance criteria. These state the specific attributes to be measured. (The performance requirements identify the problem in the context of the user's needs but do not satisfactorily identify or define the problem in terms of the attributes that should be considered in evaluating the proposed solutions.)

Performance evaluation: The third level involves the development of performance evaluation techniques. At this point in the process, the technological solutions developed in response to the user's requirements must be evaluated in terms of the established criteria. If existing test methods are not applicable, new ones must be developed. The scope of the performance concept, however, may include criteria that do not lend themselves to numerical evaluation. In such cases, simulation techniques may be required to evaluate the ability of a particular technological solution to meet the user's requirements. It is at this stage in the process that subjectively determined requirements and criteria are combined with objective measurement techniques insofar as this is possible.

Performance specifications: At the fourth level of the hierarchy are performance specifications. These are the criteria that define how the user's requirements are to be met. They also include evaluation techniques and identify the range of measurements acceptable for determining solutions. Such specifications may be used by designers, engineers, and customers seeking products, devices, processes, or systems that will serve their needs.

Performance standards: The fifth step in the process is the issuance of performance standards. In this step, standards-writing organizations may consider the specifications for publication as standards that can be referenced and used by others. By common use some specifications may become de facto national or international standards, especially if they are published by a recognized standards-writing body such as ASTM. When a performance standard is adopted by a regulatory body such as a local city council or state board and is made mandatory it becomes a performance code, which is the last step in the development of a performance test.

The process is illustrated graphically in Figure 22.1, in which one of the performance requirements for a tire is that it permits the car to be controlled in relation to the road (road holding). Road holding is measured in terms of the performance criteria, such as the capability of the tire to negotiate curves (cornering), to slow down and stop (braking), and to accelerate. Braking ability may be evaluated by

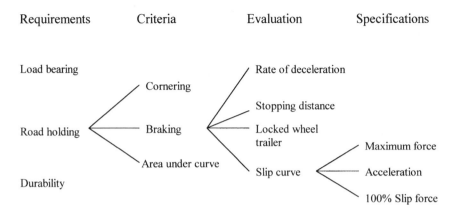

Figure 22.1 Hierarchical relation of stages in the development of performance standards.

measuring the rate of deceleration of a vehicle, measuring its stopping distance in a locked-wheel skid, measuring the braking coefficient of the tire with a trailer, or determining the slip curve from a 0 to 100% slip with some device. If we evaluate the braking ability of the tire from the full-slip curve, we may choose as the performance criteria one or more of the following attributes to be measured: the area under the curve, the maximum force, and the 100% sliding force.

Performance standards require the establishment of specific minimum (or maximum) values for each of the various components of the specifications selected in order to evaluate the criteria. In this example, braking is one of the criteria that defines the broad class of requirements called road holding. There would be a similar hierarchal arrangement for each of the performance criteria and the performance requirements.

CORRELATION

An essential step in the development of the performance test method is the determination of how well it evaluates the behavior of the product in use. In engineering jargon the question is usually put, ''How well does it correlate?'' The word *correlate* and its derivatives have a variety of meanings. From *Webster's Third International Dictionary* we have the following definitions: *Correlate*, n, 3: a phenomenon that accompanies another, visually also paralleling it (as in form, type, development, or distribution) and being related in some way to it. *Correlate*, vi, 2a: to establish a mutual or reciprocal relation of, b: to determine, establish, or show a usually causal relationship between, 3: to establish a one-to-one correspondence of (two sets or series of things): relate so that to each member of one set or series a corresponding member of another is assigned.

Correlation, n, 3: an interdependence between mathematical variables especially in statistics. *Correlation coefficient*, n, a number that serves to measure the degree of correlation between two mathematical variables.

From the *Oxford English Dictionary* come the following definitions: *Correlate*, 1: each of two things so related that the one necessarily implies or is complementary to the other, 2: more generally: each of two related things either of the terms of relation viewed in reference to the other, 3: something corresponding to analogous: analogs rare. *Correlated*, mutually or internally related.

The reader recognizes that the meanings run the gamut from mere association of two sets of terms to mathematical relationships connecting two sets of measurements. Obviously, it is necessary in any specific case under consideration to define explicitly what meaning is intended. Thus, if a performance test causes a product to fail in the same way that the product fails in service, the test correlates for failure, even though there may be no quantifiable relationship but merely a qualitative one. To draw an example from tire testing, tires that fail in a laboratory wheel test when run at higher and higher speeds fail in the same way when run on high speed tracks at increasing speeds following a certain test procedure. They fail by throwing rubber either as chunks or as large pieces of tread, and if the tests are stopped just short of throwing rubber, the tire will show separations at the shoulder or under the tread.

In associations that lend themselves to quantitative relations, we may use a statistical measure of the "quality" of the correlation; namely, the correlation coefficient. Using this measure, we would conclude that no correlation exists if a confidence interval for the population value contains the value zero for some high confidence level, such as 95%. A $+1$ or -1 coefficient means perfect correlation; a negative number means the relation is inverse, that is, when one set of values increases, the other decreases.

Rarely, if ever, do correlations based on physical measurements give correlation coefficients of unity. To illustrate such a correlation, we show in Figure 22.2 the scatter diagram for the temperature in degrees Celsius plotted against the temperature in degrees Fahrenheit. Because one set of numbers was computed from the other, the relationship is perfect and the correlation coefficient is 1.

In Figure 22.3 we show the scatter diagram for the rates of wear for the same group of tires measured on the same course at two different times. A high degree of correlation is expected since it shows the association of two sets of estimates of the same property on the same tires. The correlation coefficient is 0.936. In this case, we may interpret the scatter as chiefly due to experimental error and different environmental conditions. The scatter diagram for the estimated wear rates of duplicate tires on two different courses at different times is shown on Figure 22.4. The scatter is somewhat greater, but clearly the two data sets are associated. The correlation coefficient here is 0.821.

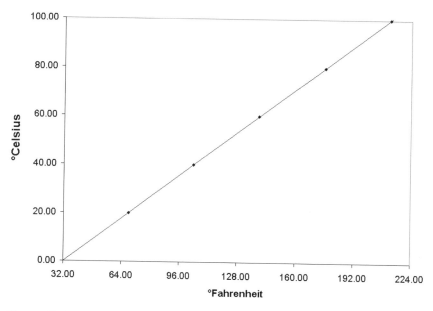

Figure 22.2   Temperature conversion from degrees Celsius to degrees Fahrenheit.

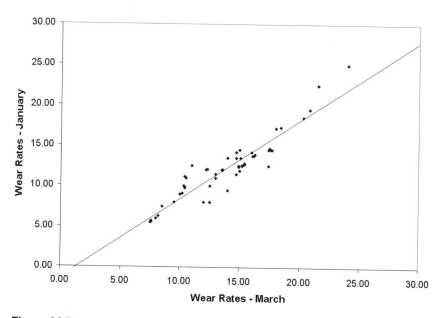

Figure 22.3   Wear rates for tires run on the same course in January and March; correlation coefficient = 0.936.

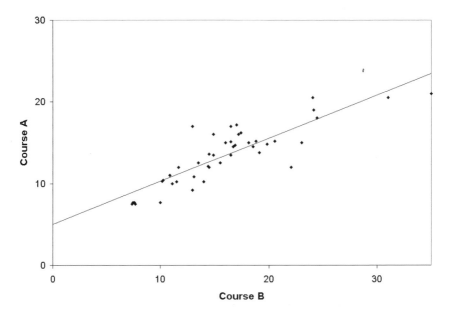

Figure 22.4   Wear rates on different courses; correlation coefficient = 0.821.

In both of the cases cited, the same property (tire-wear rate) was estimated by the same experimental technique; in one, the procedures were the same except that the determinations were made at different times, and the correlation coefficient was large. When a similar experiment was run, but on different courses, with duplicate tires instead of identical specimens, and at different times, the correlation coefficient was somewhat smaller, but some degree of correlation clearly existed. For this particular performance property and test method, these examples indicate that the greater the differences in the experimental conditions, the smaller the correlation coefficient even when there is true correlation. As additional variations in experimental conditions are introduced the coefficient may become even smaller. For example, using different experimental techniques to measure the same performance property will usually result in a smaller correlation coefficient than that found when the same evaluation technique is used.

In order to determine whether correlation exists, the prudent experimenter will run a statistical test to determine whether the correlation coefficient is significantly different from zero with some reasonable degree of confidence, such as 90, 95, or 99%. If he finds that a confidence interval for the population value at some high confidence level does not contain the value zero, he will conclude that a positive or negative correlation exists. An example of this kind of test is provided by a test of a traction property of tires. In a situation in which a driver wishes to stop as quickly as possible to avoid an accident he may apply the brakes so fast that the wheels lock. The vehicle will then slide and the frictional forces developed at the interface between tire and roadway will slow the vehicle and finally bring it to a stop. The distance the vehicle slides is dependent on its speed, the slipperiness of the road, and the properties of the tire, among other factors.

The results of "stopping distance" tests run using an automobile are variable. That is, they are not very reproducible. There are numerous factors that cannot be well controlled or compensated for and, therefore, there is relatively large experimental error. The usual way to evaluate a tire for this performance property is with an instrumented trailer or truck. One such device is a large truck on which the test tire is mounted to a special fixture equipped with a brake. Sensing equipment permits the determination of the frictional force at the tire road-surface area of contact. To make a measurement, the truck is brought to speed and held precisely at that speed, the test tire is lowered to the road, the brake is applied, and the wheel is locked for three seconds. The forces developed are then recorded. Without going into detail, the data obtained from this test device are quite reproducible. In other words, a relatively small experimental error is associated with these data.

For such data a "ranking" test is appropriate if sufficient data are available. The items tested are ordered (ranked) according to the test result from the largest to the smallest value, or vice versa. The Spearman rank correlation coefficient is determined. If that coefficient is $+1$, the same order exists in both ranks. When a number of different tires were tested on the same surface by the two procedures, the Spearman rank correlation coefficient for the two sets of results was found to be 0.50. A statistical test showed that this coefficient was not zero with 95% confidence.

The fact that the correlation coefficient is small cannot be taken as evidence that there is no relationship between the two sets of test results. Since the frictional force developed by a tire at any speed is generally considered the controlling factor in stopping a vehicle, and this phenomenon is consistent with the laws of physics, it would be surprising if significant correlation did not exist. The fact that the coefficient is small may be the result of the large experimental error associated with vehicle testing, a point that will be discussed further in the following text.

PERFORMANCE TESTS AS PREDICTORS OF PERFORMANCE IN USE

There are many reasons why performance tests conducted under controlled conditions do not predict actual performance in use, as outlined in the following sections.

Experimental Error

No measurement device is perfect. In every case in which precise measurements are made there is an error associated with each measurement. Even for situations in which the two measured properties are connected in a simple way, the correlation coefficient will be less than one. For example, the length of a piece of wire is related to its temperature; with an increase in temperature the wire expands and the length increases proportionately. If the length is measured at various temperatures, the correlation coefficient will be found to be less than unity. If the measurements are made as precisely as possible, the coefficient will closely approach unity but, because of errors in the measurement of length and temperature, it will not achieve it.

In many performance tests involving large devices the measurement error may be large. For example, the temperature of the air in a boiler or in the cavity of a tire is not uniform. Any measurement or average of several measurements cannot entirely characterize the tire's performance with respect to temperature-dependent properties because the temperature varies from place to place. In such situations, the relationship between the measured property (in this case, temperature) and the performance property will have correlation coefficients that depart significantly from the perfect value—unity. As a general rule, the larger the error or uncertainty in the measurement, the less useful are the data values for analysis.

Environmental Conditions

Laboratory tests are conducted under controlled conditions. For example, tests on tires are run under conditions of constant temperature and constant load for fixed times. In normal use, the owners of tires expose them to an infinite variety of conditions of use and misuse. Data collected under controlled conditions cannot be expected to give perfect or even near perfect correlations with observations (data) collected under the indeterminate and invariably complex and disorderly conditions prevailing in service, even when the observed properties are, in fact, correlated.

Multiple Correlation

In many cases, a dependent variable is dependent on more than one factor. In such situations, the correlation between any one factor and the dependent variable will be characterized by a small correlation coefficient. The fact that the correlation coefficient is small does not mean that correlation does not exist nor that useful information is not presented by the single factor. For example, forecasters of economic trends use several indicators, among which are machine tool orders, inventories, private debt, and housing starts. These forecasters would argue that each of these indicators predicts (correlates with) future economic behavior to some extent and therefore is useful, whereas total forecasting would depend on the complex relationships between these and other factors. To illustrate, the existence of large inventories would presage a decrease in the orders for new goods which, in turn, would be expected to result in an increase in unemployment and a general downturn in the economy. However, other factors may interfere. A tax decrease may result in an upturn in sales and the economy may not experience the downward trend.

An analogous situation occurred in tire testing in which duplicate tires were tested on a high-speed track and on a laboratory test wheel. In each, the tires were run to determine the highest speed at which they could be run without failure. The Spearman rank correlation coefficient was as low as 0.5 and yet was significantly different from zero with 95% confidence. These test results relate to the ability of a tire to operate safely at the temperature it generates while running under load. The speed at which it fails in service or on a track will depend on its operating temperature characteristics as well as on other factors. The correlation coefficient relating the failure speed on a wheel to the failure speed on a track may not be large, but the information obtained from the temperature-resistance test is useful.

CONCLUSIONS

Laboratory performance tests are designed to estimate the capacity of a product to achieve some desired performance during use. These tests are usually run under controlled conditions, whereas the product is actually used under a variety of indeterminate, disorderly, and complex situations. Also, laboratory tests frequently test for one of many factors, whereas performance in service depends on all factors.

For these reasons, and because errors of measurement always exist, a perfect relationship between laboratory test results and the performance of the product in use never exists. Statistically, the degree of correlation is measured by a correlation coefficient with unity signifying a perfect correlation. When the coefficient is not large and there is doubt as to whether or not correlation exists, a statistical test is made to determine whether the true correlation coefficient could be zero. If it appears that the true coefficient is not zero with a high degree of certainty, and if the laboratory and in-service test procedures have a common physical basis, then some correlation may be assumed to exist.

REFERENCES

1. GJ Rothwell. The Performance Concept: A Basis for Standards Development. Vectors 2, 1969.

23

Standardization and Quality Assurance in Developing Countries

This chapter has been prepared by Balbir Bhagowalia of Sterling, VA. The work is based on his experiences as a quality-assurance expert under contract with the United Nations Industrial Development Organization (UNIDO).

INTERNATIONAL TRADE AND THE NEED FOR QUALITY AND PRODUCTIVITY DEVELOPMENT

With the recognition of the importance and benefits of globalized trade, the interest in industrial development in developing countries has been growing. Various aspects of international trade are discussed in the media. Many discussions compare the social, political, economical, technological, and cultural aspects of industry and trade in these countries with those of the industrially developed nations. One of the important aspects that is overlooked is that the achievement of economical and quality production are important for industrial growth for developing countries and for the enhancement of trade with the industrialized markets.

Among the essential elements that are required are the establishment of standardization and quality-assurance management practices, which are necessary for developing countries to achieve industrialization and strengthen international competitiveness; and to improve quality and productivity through rationalized production. In addition, they are also prerequisites to promote smooth transfer of technology from the industrialized to the developing countries. Therefore, they constitute the area of strategic importance on which emphasis must be placed to assist developing countries promoting their industrial development.

However, there is still not enough awareness in developing countries of the significance and importance of these elements upon their industrialization efforts. This chapter of the book discusses and describes some of the important aspects of these activities to indicate the salient differences in the utilization of these activities in developing countries for quality and productivity improvement in industry—particularly the engineering industry. It also discusses the needs and methods of providing assistance to these countries to achieve industrial growth through use of these activities.

APPROACH TO STANDARDIZATION ACTIVITY IN DEVELOPING COUNTRIES

Organized standardization evolved in industrialized countries. In most of them, standardization was practiced at company or industry (association) levels, even before national standardization. In developing countries the case is different. Most of them initiated national standards activity as a part of their national development plans and from that point on some of them built it up relatively rapidly. However, the activity has not percolated down to industry or company levels at the same pace. In addition, in developing countries the industry has grown mostly on imported know-how, design, technology, equipment, materials, and components from industrialized countries. Consequently, in the initial stages there is considerable dependence on imported, industrialized country standards or designs. As such, few efforts are made in these countries to have organized company or industry standardization activity. It is important to see how this situation affects standardization and quality related activities at various levels and to see what influence it has on the development of industry.

STANDARDIZATION AT NATIONAL AND COMPANY LEVELS IN DEVELOPING COUNTRIES

Differences in Emphasis

While objectives of the standardization activities in developing and industrialized countries are similar, there are certain areas that need more emphasis in developing countries to meet the special situation created by imports and imported technology. These are:

1. Promotion of standardization and quality assurance (QA) at industry and company levels;
2. Aggressive implementation of standards and QA techniques;

3. Encouragement in indigenous (local) development of industrial production;
4. Emphasis on import substitution and use of indigenous materials;
5. Assistance in export promotion;
6. Increased attempts in variety reduction of materials and products;
7. Means to ensure optimum quality of essential goods and services; and
8. Ensure standards take care of local conditions.

In addition to identifying the activities on which more emphasis is required, the national standards bodies in the developing countries have to give guidance and assistance to the industry and occasionally undertake tasks of fulfilling these aims. In industrialized countries, many of these tasks do not need emphasis because they are routinely performed by industry. Some examples of the types of work that need to be attempted in developing countries will illustrate this point.

Variety Reduction

Companies in most developing countries have collaboration agreements with companies in various countries that include technical and business know-how, plant, equipment and materials, including standards and designs. Consequently, the production in such developing countries were overwhelmed with hordes of different national, association, and company standards as well as different designs or multiple brand names. A situation may develop where for a single item, numerous varieties (sometimes differing only slightly from each other) get used in small quantities. This leads invariably to stocking of unnecessarily high inventories, uneconomical production, slow development of indigenous items, and general confusion. Solutions to such problems require various corrective actions, one of the most important being organized variety reduction or rationalization programs.

For example, in India a similar situation occurred some years back when it was found during a study that over 1,500 alloy and special steel specifications were used in engineering industry. While some quantities required were large enough for economical production or import, others were so small that the importing and storage costs were many times more than the price of the items (because of small quantity requirements, their indigenous development was not economical). The Bureau of Indian Standards (BIS) took up the task of rationalization of these steels and succeeded in reducing them, in successive phases, to 130 types and aligning them with international standards. Later efforts were made in specific areas, for example, in the automobile industry to restrict use to even fewer steel specifications, thus increasing the quantity requirement for each specification or size. To appreciate such drastic measures one has to realize that in developing countries, even the size of India, production of some items is relatively low and only a reduced variety will ensure larger economic quantities and facilitate start of manufacture of these materials.

Another example is that of a steel plant, where gear boxes for different applications were imported and attempts to develop them indigenously failed due to there being too many types, each in small quantities. The company standards department rationalized them into fewer types after a detailed study and got the other steel plants interested in using these standardization techniques for variety reduction. The result yielded savings, import substitution, indigenous development, quicker deliveries, reduced down time, and other benefits.

Quality Assurance Services

For ensuring that quality goods and services be provided, a tradition of QA and total quality management (TQM) has to be built up. In industrialized countries this has evolved slowly over the years, from inspection to quality control (QC), statistical quality control (SQC), quality assurance (QA), total quality control (TQC), TQM, and ISO 9000 certification, and national quality awards. Developing countries must, however, make concerted efforts to achieve this goal more quickly if they wish to establish a proper, economically viable industrial structure and wish to participate in world trade. To promote industrial quality at the national level in the developing countries (at least in the initial stages), the national standards bodies have to take major responsibility for propagating QA activities.

In many developing countries, the national standards bodies are taking on these tasks. In India for example, BIS offered SQC consultancy services and training programs at nominal charges for improving quality and reliability of the goods and services and cutting down costs of rejections and rework by establishing QA and SQC systems in companies. In some other countries initiating industrialization, the national standards bodies (which are mostly government bodies) even give free QA services, including information on typical QA systems for common product groups.

Meeting Local Conditions

In many cases, direct adoption of foreign national standards in the developing countries can cause some operational, quality, or economical problems due to differing local conditions in respect to usage, methods of manufacture, harsher environment, etc. In such cases, the national standards body of the developing country has to review the foreign standards and adapt them with suitable additions or amendments.

Company Standardization Services

Benefits from national standards cannot be optimized or achieved unless the standards are implemented. It has been found (from experience in India, Brazil, and some other countries) that in most cases companies with in-plant standardization activity accommodate more readily to national standards and their adoption. However, there is not enough company standardization activity unless the national standards bodies in developing countries become active spokesmen to promote company standardization and train personnel. For example, in India BIS was instrumental in running many seminars and training courses, and giving individual consultancies to companies and then encouraging the latter to share their experiences with others. In Brazil, the national standards system held many seminars with companies in important industrial cities on the subject of standardization in general and company standardization in particular. In India, the Institute of Standards Engineers (SEI) propagation of company standardization was founded with active help from BIS.

In developing countries, the national standards bodies must make concerted efforts to get standards adopted directly by government departments, government purchasing agencies, large undertakings, and small industries by establishing departments responsible for purposes of standards implementation and company standards promotion. This will normally encourage industry to adopt standards.

Import Substitution and Indigenous Development

Indigenous development of certain items is not possible in developing countries because of the small quantities required or because lack of manufacturing information on the items being imported or information on locally produced items that could have been used. Efforts by companies in the developing countries is not normally sufficient to identify such substitutes, but efforts through national standardization committees in cooperation with national research labs can assist in such substitution.

Other Problems

Some of the foregoing examples show that in some areas, the companies in developing countries have to face complex situations due to multiplicity of standards and imports, high inventories and lengthy storage of materials, delays in imports, and lack of suitable data in some fields, etc. Some problems are more or less peculiar to developing countries. Various examples from such countries show that formalized and organized company standardization have helped in solving quality and productivity problems.

NEED FOR ASSISTING DEVELOPING COUNTRIES IN STANDARDIZATION AND QUALITY ASSURANCE

Ultimately, the developing nations will have to develop more of these programs by themselves and adopt means of utilizing QA techniques and standardization to ensure faster industrial growth, economic production, quality and reliable products and services, enhancement of exports, and less dependence on imports. This is being done in many countries through the establishment of national standards bodies and related infrastructure to assist in the quality and productivity improvements that will augment other means used to have a growing and competitive national industrial sector. However, these nations face problems such as lack of funds for standardization work, shortage of trained personnel in QA and standardization fields, shortage of standards, insufficient involvement in international standardization work, and consequent difficulties faced in adopting such standards. There is also a lack of national experience demonstrating benefits from organized and formalized standardization and QA. This means that in the initial stages, most nations need organized and formalized external assistance in the area of standardization and QA to enable them to start in a direction that allows them to achieve their development objectives.

Assistance is being provided to the developing nations in many forms of technical help, transfer of technology, and standards educational and training programs in both standardization and QA-related areas. Such programs are sponsored by agreements between international bodies (such as the United Nations) or industrialized countries' aid agencies (e.g., U.S. Agency for International Development USAID) and the government of a developing nation. Most of the assistance in training in QA techniques, standardization, or in the formulation and implementation or application of standards areas is provided through experienced standardizers to counterparts in developing nations. In many cases, such transfer of knowledge had been successful and resulted in the initiation or improvement of organized QA and formalized standardization activity; in building up institutional infrastructure for carrying out standardization and QA work; in major benefits to economy and quality; and in training of local personnel in principles and techniques of standardization.

The areas of assistance have been diverse and cannot be classified into a few types. However, some of the major categories are:

1. Setting up of QA/QC services
2. Introduction or improvement of certification/quality-mark systems
3. Setting up of laboratories for testing
4. Setting up of consumer protection systems
5. Improvements in standardization propagation
6. Setting up of modern technical-information systems
7. Introduction of concept of organized company standardization
8. Initiating rationalization of products and materials
9. Integration of standardization and QA in specific areas (such as automotive components, etc.)
10. Improvement in standards elaboration systems

11. Enhancing means for implementation of standards
12. Facilitating metrication
13. Enhancing export promotion and preshipment inspection
14. Setting up of metrology systems
15. Establishing means for calculating economic benefits from standardization and QA activities
16. Training in areas of standardization

METHODS OF PROVIDING EFFECTIVE ASSISTANCE

An Integrated Standardization and Quality Assurance Program

For programs in quality and productivity improvements it is necessary to have integrated programs in standardization and QA. An example of such a program carried out in a developing country under the auspices of a technical assistance program sponsored by an international development agency is described in the following text. Such programs are conducted in various countries in different industries under sponsorship of both international agencies (such as the United Nations Development Programme, UNDP) or agencies from industrialized countries (e.g., USAID) assisting in industrial development of the Third World. Although no specific countries and agencies are named, the examples are taken from actual programs.

In one example, programs were instituted to assist an automobile component industry in developing so that it can grow by supplying components to original equipment manufacturers (OEMs) (who were importing components from other countries); by supplying the replacement market of industrialized countries with spare parts (which at present is dominated by imports); and by initiating exports of some parts or products to industrialized countries as well as to other developing countries. Two purposes are served by this action: the developing country receives foreign exchange and they also obtain equipment, materials, and know-how. Earlier, vehicles were assembled mostly from imported parts, but the government and industry have been trying to increase the use of locally produced components. The program has been only partially successful in light commercial vehicles, whereas in other vehicles this content is very low. In terms of number of components and local content of parts and raw materials for making these components, the localized portion is even lower.

There are many socioeconomic and other reasons for this situation (such as the large variety of vehicles) but there are also technical reasons that contribute to this adverse situation. The requirements of the vehicle assemblers (OEM) in respect of performance, reliability, cost, and delivery of components cannot be met by the indigenous component manufacturers unless they have means to achieve higher and consistent quality and economical production. The failure to apply standardization and QA techniques at national, industrial, and company levels is a major cause for not meeting OEM requirements or not providing quality, reliable, and economical spare (or after market) parts.

This problem was examined and the possibility proposed of providing assistance through an international development agency's project plan that would ensure a beneficial and long-term solution to the automotive component industry.

Project Strategy

The strategy proposed and implemented later was through a two-year project described in the following text. The strategy for the project was to demonstrate that through organized application of standardization and QA techniques, the quality of goods and economics of production can be upgraded. The industry can then graduate from the prevailing situation, where only a few items were supplied to the OEM and replacement markets, to a situation in the future where a bigger share of the replacement market and supply to OEM is taken by the local industry (small- and medium-size enterprises). Consumers will benefit also.

This possibility of achieving quality production economically was to be demonstrated by the project through selection of few enterprises and to upgrade their production and quality systems to enable them to either supply to OEM or get the national certification mark through improved quality. This work by the project was to be done through provision of more and better national and industry standards and their implementation in industry through certification and other means and application of quality systems. Simultaneously, with this approach focused on selected components and manufacturers, the bodies responsible for drafting standards and for implementing certification systems, including testing, were to be trained and strengthened in order to ensure sustainability of the initiated activities.

The work was to be done by local staff formed in a core team who were to be assisted by the project through:

1. Providing intensive theoretical and practical on-job training by international consultants in standardization, quality management techniques, and testing;
2. Training the core team who will then train others to apply the techniques demonstrated;
3. Introducing some new activities such as rationalization;
4. Improving procedures, practices, and activities in standardization and QA (especially in standards making and implementation);
5. Providing some testing equipment, accessories, and supporting information for testing of automotive components; and
6. Using national experts in selected areas in the training programs.

This model, when found successful, could then be extended to other automotive and nonautomotive sectors, and later, through international agency support, to other projects.

Planning, Design, and Activities of the Project

As seen from the foregoing, the project did not envisage long-term technical assistance in various technology areas, but it aimed at identifying causes that prevent the smaller, local manufacturers from supplying components in the correct quality at competitive prices and on time; and then suggesting means to overcome the identified shortcomings.

Once the situation was clear, a few automotive components and the smaller units manufacturing them would be selected from those surveyed to initiate more detailed work to identify problems causing low productivity and quality, and demonstrate means to overcome them. Simultaneously, work was to be initiated in standardization-related organizations to identify problems with standards development and implementation, especially with the standards on the components. Means were later to be indicated to make improvements in this area.

Work was also to be initiated in the laboratories and testing-related institutions and industry to identify shortcomings in the area of testing of automotive components and showing means to ensure improved testing of components through effective test methods and standards, suitable test equipment, better trained staff, application of QA in laboratories, and providing an advisory service on how to meet requisite quality in components.

These tasks were to be performed normally by members of a ''core team'' selected from participating industry, standards organizations, and the laboratories. The core team members were to work in conjunction with staff from the participating organizations under guidance of international and national experts in different areas such as standardization, QA, testing, and certification. The activities of the team would produce outputs that could be utilized in various ways.

1. For training of core team and others through training programs and workshops
2. For promotion of standardization and QA techniques through seminars
3. For upgrading production in smaller enterprises
4. For documentation (QA manuals, case studies, codes and practices, systems, procedures, effective standards, etc.)
5. For reports by experts and counterparts (recommending action by the industry, government, industry associations)

The main target beneficiaries were to be the small- and medium-scale industries in the country who would learn from the project activities and results as to how to attain required quality and reliability in their products and how to improve productivity in their operations.

PROJECT RESULTS

Meet Immediate and Overall Objectives

As indicated earlier in this case study, the development objective of the project was to promote productivity and to assist the industry in that country to mature growth through producing goods with acceptable quality and economically by application of standardization and QA techniques. This would also ensure acceptability of products in the spare parts market and, consequently, lower imports as well as develop export potential. Initially, the techniques would be applied in small- and medium-scale industry engaged in production of automotive components and parts. Model projects developed and successfully applied could later on be extended to other industrial sectors.

The immediate objective was to assist in the overall aim through increasing and accelerating the pace of implementation of standards that would help in productivity and quality improvement in the automotive industry. The major elements of the project that were met are:

1. Strengthening the capability of the national standards body;
2. Enhancing the level of implementation of the national and company standards;
3. Improving product and quality system certification marking of automotive components;
4. Strengthening and expanding the capability of the domestic testing laboratories;
5. Initiating efforts to laboratory accreditation internationally;
6. Upgrading the capabilities for quality and economical of production in selected units;
7. Initiating concept of rationalization to achieve economy and quality; and
8. Enhancing knowledge in application of standardization and QA in industry.

Utilization of Project Results

The probability of utilizing the results from the project activities could be judged from the positive actions taken by participants during the project and commitments shown by senior management in all sectors when their plans indicate that they would carry on follow-up actions on the recommendations made by the project team (see Figure 23.1).

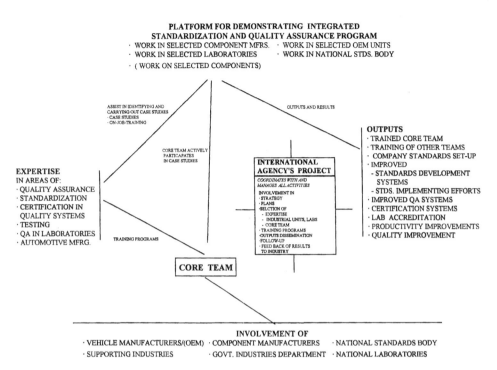

Figure 23.1 Project design and methodology for an integrated standardization and quality assurance program.

This chapter has described some of the reasons and methods for formalized assistance to developing nations from international organizations and industrialized countries. In the final analysis it can be said that with increased industrialization and expanding international trade in the world, standardization and QA are of critical importance to develop and strengthen the industrial technology base in developing countries to ensure free trade and avoid further barriers to cross-border movement of goods and services.

REFERENCES

1. BS Bhagowalia, JL Donaldson, RB Toth, and SM Spivak. Standards Education, Part III: Special Needs of the Developing Nations. Standardization News 16(3):54–60, March 1988.
2. BS Bhagowalia and SM Spivak. Standards Education: A Critique of the Assistance Provided to Developing Nations. Unpublished article.

24

International Standards: The Key to Global Competitiveness

This chapter was contributed by Donald R. Mackay, Director, International Standards, Air-Conditioning and Refrigeration Institute (ARI), Arlington, VA and a Fellow and past President of the Standards Engineering Society (SES) of North America. The chapter describes the international standards program of a typical American trade association and explains how industry standards form the basis for the development of international standards. The standards-development activities of several international subcommittees are described, illustrating the initiatives taken by American industry. The industry philosophy that global competitiveness will be based on internationally agreed standards is explained and promoted.

INTRODUCTION

The ARI is a trade association representing 200 companies manufacturing over 90% of the central air-conditioning and commercial refrigeration equipment produced in the United States. ARI was established in 1953 with the merger of the Refrigeration Equipment Manufacturers Association and the Air-Conditioning and Refrigeration Machinery Association. In 1965, the members of the National Warm Air Heating and Air-Conditioning Association joined ARI, and in 1967, the members of the Air Filter Institute joined ARI.

The primary objectives of ARI are (1) to develop and implement voluntary product standards relating to the performance of the products manufactured by the members of ARI, (2) to establish and maintain certification programs for products manufactured and sold as conforming to those voluntary product-performance standards, (3) to promote the use of ARI standards and certification programs as the basis for industry-wide competitiveness in the U.S. marketplace, and (4) to promote the initiation and development of international standards in International Organization for Standardization (ISO) and International Electrotechnical Commission (IEC) that can be used to facilitate competitiveness in the global marketplace.

ARI's secondary objectives include (1) the provision of knowledge and information to governmental regulatory and procurement agencies, and international trade organizations to facilitate mandated programs, (2) the development and distribution of information to the public in the form of consumer brochures, (3) the education and training of technicians who install, service, and repair air-conditioning and refrigeration equipment, (4) assisting industry in international trade activities, (5) providing detailed statistical information on industry product shipments, and (6) advancing industry technology through cooperative research programs.

ARI is composed of 200 companies that produce central air-conditioning and commercial refrigeration systems as well as companies that produce associated equipment and components. These companies are organized into ''Product Sections'' that are responsible for the technical and political activities related to the products within their scopes. Each Product Section has an Engineering Committee responsible for the development of voluntary standards, and a Certification or Compliance Committee responsible for the development and implementation of voluntary product-certification programs.

The ARI Product Sections meet twice a year; once in the Spring and once in the Fall. The Engineering and Compliance Committees meet in between meetings of the Product Sections as necessary to carry out the decisions of the sections. Task Groups are often formed to expedite the initial development of standards and certification programs. Meetings concerning the development of ARI product standards are open to any interested party and notice of such meetings is provided through the monthly ARI publication, *KOLDFAX*.

To date, sixty voluntary standards have been developed, approved, and published by ARI. These standards cover such products as air-conditioners and heat pumps, condensing units, packaged terminal equipment, water-chilling equipment, unit coolers, air-distribution equipment, and humidifiers. Standards have also been developed concerning the sound rating of air-conditioning equipment, specifications for fluorocarbon refrigerants, and refrigerant recovery/recycling equipment. The ARI General Standards Committee acts for the Board of Directors in grating final approval of standards.

ARI standards, once published, are submitted for approval to the American National Standards Institute (ANSI) under the Accredited Canvass Method. The standards are subjected to a 60-day public review period during which any interested party can submit written comments on the standards. ARI must respond to such comments and attempt to resolve any substantive negative comments or objections to the approval of the standards as American National Standards. At the present time, thirty ARI standards have been approved as American National Standards and ten others are in the process of being approved. Ultimately, all ARI standards will be submitted to ANSI for approval.

ARI standards generally establish methods for testing and rating the performance of equipment. ARI standards are used as the basis for product-certification programs under which the manufacturers certify the performance ratings of the products and ARI, through the selection and testing of random samples, verifies the accuracy of the ratings. The testing is carried out by an independent laboratory under contract to ARI. Approximately 30% of each manufacturer's models are tested each year.

The ARI certification programs are open to ARI members and U.S.-based nonmember manufacturers. The overall policies and procedures for ARI certification programs are established by the Certification Programs and Policy Committee. Each Product Section adds certification policies and procedures that are specific to their needs. The results of the certification process are published in ARI Certification Directories that list each model of every manufacturer in the program and the results of the performance testing—the ratings of the characteristics specified in the ARI standard.

These ARI Certification Directories are used by contractors, engineers, and architects in selecting equipment for specific installations. The certified data may include heating and/or cooling capacities, power consumption, energy efficiencies, and coefficients of performance at various standard-rating conditions. These ratings are certified by the manufacturer as being accurate under the conditions specified in the ARI standard. These ratings are accepted by engineers as providing accurate, useful information related to the performance characteristics of the products certified. These ratings have also been accepted by utility companies as the basis for providing rebates for energy-efficient equipment and by the U.S. Department of Energy in determining compliance with the Federal Energy Act.

ARI POLICY REGARDING INTERNATIONAL STANDARDS

In 1989, the ARI Board of Directors adopted a policy to support the development of international standards through ISO and IEC and to participate aggressively in the initiation and development of international standards for those products that were marketed internationally. In addition, the board adopted a policy directing Product Sections to develop, within one year of the publication of an international standard, a plan for adopting that international standard or to justify to the board the nonadoption of that standard. The board also established within ARI a full-time position for a Manager of International Standards.

This proactive policy regarding the development and use of international standards is a reflection of the corporate thinking of many of the ARI company members who have established markets for their products in other countries throughout the world. Several ARI member companies have manufacturing facilities abroad, some in as many as six countries, as well as manufacturing license agreements with foreign companies to produce air-conditioning and refrigeration equipment in other countries.

ARI is fortunate to have forward-thinking executives on its Board of Directors who believe that the development of international standards is the key to global competitiveness. Obviously, it is not efficient or economical to produce different models of the same basic equipment to meet different national requirements in order for products to be sold in different countries.

The American industry has enjoyed one advantage regarding the development of international standards that is not present in many other industries. "Air-conditioning" was invented in the United States by Willis Carrier in the early 1900s and American standards for air-conditioning and refrigeration equipment were accepted as de facto standards in many parts of the world because of the "Made-in-America" equipment shipped overseas.

ARI Involvement with ISO Activities

ARI staff is responsible for carrying out the following activities:

1. Preparing requests for "New Work Items;"
2. Organizing Working Groups with U.S. Conveners;
3. Preparing Proposed Working Draft documents;
4. Managing the development of proposed standards;
5. Preparing Committee Drafts and Draft International Standards; and
6. Responding to technical comments received during balloting.

All of these functions are carried out in the two subcommittees of most interest to ARI members: ISO Technical Committee TC86, Subcommittee SC6 (commonly referred to as "TC86/SC6") on "Air-Conditioners and Heat Pumps" and ISO TC86 Subcommittee SC8 on "Refrigerants and Lubricants."

ISO TC86/SC6, Air Conditioners, and Heat Pumps

ARI holds the Secretariat for Subcommittee SC6 on "Air Conditioners and Heat Pumps"—the most active subcommittee within TC86. Presently, there are seven working groups (WG) in SC6.

1. WG1 on Unitary Equipment
2. WG2 on Sound Ratings of Equipment
3. WG3 on Water-Source Heat Pumps
4. WG4 on Air-Conditioning Condensing Units
5. WG5 on Multi-Split Systems
6. WG6 on Automatic Integrated Controls
7. WG7 on Room Fan-Coil Units

Current Status of ISO TC86 Standardization Projects

After several years of active involvement in the development of standards in ISO TC86, the following accomplishments can be reported.

1. A revision of the ISO Standard 5149 covering mechanical safety has been initiated in a newly formed working group in SC1 on "Safety" and the first meeting of the working group was held in Paris in July 1997. The revision will involve the establishment of safety classifications of refrigerants for flammability and toxicity, as presently contained in ASHRAE Standard 34 "Refrigerant Designations," which is widely recognized by refrigerant experts around the world.

2. A draft document has been developed for SC2 on "Terms and Definitions." This document contains over five hundred terms and definitions covering many types of products within the scope of TC86.

3. Two basic standards relating to the testing of the performance of refrigerant compressors have been revised in SC4 on "Refrigerant Compressors": ISO 917, "Testing of Refrigerant Compressors" and ISO 9309, "Presentation of Performance Data for Refrigerant Compressors." These standards have been balloted as Draft International Standards (DIS).

4. Two standards covering unitary air-conditioners and heat pump equipment have been published by ISO: ISO 5151, "Non-Ducted Air-Conditioners and Heat Pumps—Testing and Rating of Performance" (includes room air-conditioners) and ISO 13253, "Ducted Air-Conditioners and Air-to-Air Heat Pumps—Testing and Rating for Performance." These are currently being revised in SC6/WG1.

5. Two standards have been approved concerning the sound rating of air-conditioning equipment in SC6: one covers outdoor equipment and the second covers nonducted indoor equipment. These standards are being balloted as Final Draft International Standard (FDIS) documents. A third standard covering the sound rating of indoor ducted equipment is being developed in SC6/WG2.

6. Two standards covering water-source heat pumps have been approved in SC6/WG3. One standard covers water-to-air and brine-to-air equipment and the second covers water-to-water and brine-to-water equipment. Both standards are being balloted as FDIS.

7. A standard concerning the testing and rating of air-conditioner condensing units is under development in SC6/WG4. The first meeting of the working group was held in London in March 1995. The document was approved as a Committee Draft (CD), but has recently been revised to cover condensing units for refrigeration equipment.

8. A new project has been initiated to establish performance testing and rating criteria for air-conditioners having one or more compressors and one or more evaporators. These units, referred to as "multi-splits" are becoming very popular because they are ductless and can provide cooling of individual rooms in a structure.

9. Another project has been initiated to establish requirements for automatic integrated controls used in air-conditioning systems.

10. A new project has been established to develop an ISO Standard for room fan-coil units.

11. In SC8 on "Refrigerants and Lubricants," a working group has developed a document entitled, "Fluorocarbon Refrigerant Specifications and Test Methods." This standard being balloted as an FDIS establishes the technical characteristics and analytical test methods for determining the purity of the many fluorocarbon refrigerants being produced commercially for refrigeration applications

12. Another project in SC8 concerns the revision and extension of ISO 817 "Refrigerants—Number Designation," a standard that establishes the mechanics of identifying the many chemical products being used as refrigerants around the world and assigns individual identifiers to them. The "R" numbered and lettered designations provide a shorthand identification scheme for indicating the specific chemical composition of all commercially viable refrigerants.

13. A new standard produced by SC8 covering equipment used for the recovery and recycling of refrigerants is about to be published by ISO. It will be identified as ISO 11650, "Refrigerant Recovery and Recycling Equipment—Testing and Rating of Performance." A revision of this important standard is being prepared.

14. A new project has been initiated in an SC8 working group concerning the reuse of refrigerants. This project is very important because of the production bans on chlorofluorocarbon (CFC) refrigerants and the phaseout of the HCFC refrigerants. The proposed standard will establish maximum acceptable contaminant levels for the reuse of recycled refrigerants.

ARI Involvement in IEC Activities

ARI has, in addition to being involved in the development of testing and rating of equipment for performance, been concerned about the development of requirements and test methods concerning the evaluation of electrical safety. Within the

IEC, Technical Committee TC61 is responsible for standards relating to domestic appliances. Subcommittee SC61D is specifically concerned with electrical safety requirements pertaining to air-conditioning equipment.

The basic document for TC61 has been established as IEC 335, "Safety of Household and Similar Electrical Appliances—Part 1: General Requirements." SC61D had previously developed a Part 2 document, identified as IEC 335-2-40, entitled "Particular Requirements for Electrical Heat Pumps, Air-Conditioners and Dehumidifiers."

In 1989, with ARI's board-directed policy for increased participation in the development of international standards, the United States sought and obtained the chairmanship of SC61D and promoted U.S. responsibility for the Secretariat of SC61D. Herb Phillips, then Vice President of Engineering for ARI became the SC61D Chairman and Paul Kolb of ETL, now Intertek, became the SC61D Secretariat. With the United States in charge of the standards-development program in SC61D, the level of activity increased significantly.

Current Status of IEC Standardization

The following projects are currently under development within IEC SC61D:

1. A revision of IEC 335-2-40 to align the Part 2 document with the third edition of IEC 335-1 has recently been published. The previous edition of this standard has been accepted by the Standards Council of Canada as an "alternate" Canadian National Standard. Efforts to adopt the latest version have been initiated. Mexico is considering the adoption of this new IEC version as a voluntary standard, to be considered subsequently as a mandatory Mexican standard or NOM.

2. Another revision of IEC 335-2-40 is under development by the industry, with additions and deviations to comply with the National Electrical Code and American Society of Heating, Refrigeration and Air Conditioning Engineers (ASHRAE) Standard 15 on Safety, to provide an alternative to Underwriters Laboratories (UL) 1995 on "Heating and Cooling Equipment."

3. A new IEC standard covering the electrical safety requirements for humidifiers to be used with air-conditioning equipment has recently been published. This standard, based on the binational UL/Canadian Standards Association (CSA) standard UL 1995 was initiated in 1993, and published as IEC 6035-2-88 in 1997.

4. A new working group has been established in SC61D to develop electrical safety requirements for refrigerant recovery and recycling equipment. Initiated by the United States, the preliminary working draft is based on UL 1963, "Refrigerant Recovery/Recycling Equipment."

5. Another new SC61D working group has been established to develop electrical safety requirements for equipment using flammable refrigerants. Con-

vened by Sweden, this working group was deemed important because of the increasing use of environmentally compatible flammable refrigerants in Europe.

6.   Within SC61D, a revision is being prepared to incorporate safety requirements for room fan-coils in the basic 335-2-40 standard. Italy is spearheading this effort.

SUMMARY

The leaders of the air-conditioning industry in the United States realized in early 1989 that the key to global competitiveness, and indeed perhaps the route to survival in a highly competitive industry, was through the initiation, development, and implementation of international standards. They realized that it was more efficient and economical to manufacture basic products conforming to internationally recognized requirements using internationally agreed test methods than to manufacturer specific models for different countries.

The ARI decision to become aggressively involved in the development of international standards, while costly in terms of time and effort, will, in the view of the author of this paper, begin to pay dividends in the next millennium. The work of initiating and developing international standards in both ISO and IEC has resulted in the establishment of seventeen individual international standards and the revision of seven existing standards and will come to fruition as these standards are published by ISO and IEC and adopted by the member bodies of ISO and IEC.

25

The Informal Versus the Formal Standards Development Process: Myth and Reality

This chapter was contributed by Carl Cargill, Director of Standards with Sun Microsystems and formerly standards strategist for Sun, Digital, and Netscape. His expertise focuses on standards development by consortia and alliances, rather than either de facto standards created by the marketplace or the traditional standards-development route. Examples include the World Wide Web Consortium (W3C) and The Internet Engineering Task Force (IETF). This approach is a more pointed argument for consortia than, for example, that described by C. Ronald Simpkins for the chemical process industries in Chapter 19. Carl Cargill is author of *Information Technology Standardization* and an expert on information technology standardization and a proponent of the consortia standards process.

SUMMARY

The informal standardization development process consists of two primary areas, consortia and alliances, and de facto or marketplace-driven standardization. This chapter will focus on the first of these phenomena, the consortia and alliance process, since describing the de facto standards process is very difficult—due to the nature of the beast. The problem comes about because there are no good descriptions of how a de facto process can be made to work. While there are numerous academic articles on how such standards impact the market, or how they "theoretically" can be created, there is no good descriptor of the process by which an organization can create a de facto standard. The creation of a de facto standard is based upon market forces such as product availability, skillful marketing, and timing—three attributes that contain an infinite number of attributes and are subject to speculation. I will, therefore, focus on the areas of consortia and alliances as the primary manifestations of the informal process.

CONSORTIA—THE RATIONALE

Consortia have often been portrayed as an aberrant form of standardization organization, one that results from people not understanding the strength and goodness of the formal standards developing organizations (SDOs). This section looks at the myths that surround consortia to show why consortia are so successful in highly dynamic markets such as Information Technology (IT).

The SDOs, which have been explored in detail in earlier sections of this book, represent a formal and structured way of achieving consensus on things that impact millions of people. They provide a degree of stability that helps both users and providers clarify and respond to market needs. By acting as the voice of impartiality, the SDO can lay claim to being the rational force in the market, creating standards that serve the users and the producers. They save the best of the past, merge it with the best of the future, and encourage providers to make better products and work together.

Within the IT arena, however, this describes the mystic ideal, an ideal that is a myth and that falls far short of reality. The reality is that SDOs are creatures of the vendors. Nearly all participants in the formal process are there at the behest of their employers, who pay their salaries, travel costs, and expenses, and who provide the participants the right to take time from their jobs to attend these meetings. Vendors send representatives to these meetings for a business purpose. Very few vendors are altruistic about providing support to these organizations. The whole SDO arena is charged with the need to protect one's organizational interests. This protection of an organizational interest has twisted and changed the nature of the formal process within the IT market.

The reason for this behavior is not hard to find. Within the IT community standards are a business issue. The IT industry was a bellwether industry in making standards a part of the business process by stressing that standards would be a named product attribute in the design and build phases, and then in the procurement of product. While other industries used standards as part of their product cycle, it was only in the IT industry that standards were used as a product differentiator. As early as 1986, there was an undercurrent within the industry that was focused on basing standards on business considerations. The idea surfaced in the 1988 time frame as a new form of rationale for participation in standardization activities (1). The idea was picked up quickly by the IT industry, which was in the midst of a looming battle over "open systems" and what that meant. The concept quickly spread through IT standardization players and then spread more widely—in the literature of the day. By the late 1980s the idea was reasonably well known in the literature on the discipline of standardization (2).

However, this idea was most firmly established in the IT community. Again, the reason for the IT industry being this receptive to the siren song of standardization was that the nature of business in the IT sector was changing. Some background on the nature of this change is necessary.

In the 1960s, a major battle about "plug compatibility" began. Basically, plug compatibility permits devices from a multitude of different vendors to be used on the same computer system. The reason for this was that the dominant supplier of mainframe computers, IBM, charged what many in the market thought were high prices for their rotating disk memory, and these users wanted to be able to plug in another vendor's disks. This, of course, required standards and marked the start of standardization in the IT industry. This continued through data communications, modems, plugs, languages, and finally into operating systems (POSIX) (3). The vendors were responding to user needs for standards and the users were repeating back to the vendors the things they were reading in the popular press that had been largely derived from sources among the providers.

This finally worked within the hardware area. Keyboards and keyboard plugs, video attachments, communications attachments, disk attachments—all became standardized. Within the software environment, however, while demanding standards and standardization, the users continued to reward vendors who were nonstandard. In a particularly strange twist of fate, the users—while voicing support for standards—were rewarding providers of newer, faster, and more interesting nonstandardized products and services. This left vendors who were providing standardized solutions in the unenviable position of having higher product costs (since standardized interfaces are more expensive to build and operate) and less market appeal. The lesson was not lost on the providers, who then began to try to understand how to solve this—and other—standardization problems.

Within the industry, the first goal was to evolve the de facto standard. This standard promised a single implementation of a specification to ensure interoperability and spread the acceptance and dissemination of the technology, which would lead to market dominance and economic payback that was desired. There were many attempts to accomplish this—most of them unsuccessful. Of the ones that did succeed, however, some were very successful. The maintenance and growth of a de facto standard, however, is difficult and requires concentration on the dynamics of the market. Organizations that succeed with de facto standards over the long term are usually competent marketing firms with technical capability.

If the players in an organization couldn't cause a de facto standard to emerge, the second drive was to get a standardized specification out faster than the SDOs could—hence the consortia. The consortium was a collection of like-minded companies who were devoted to doing something using the same basic technology and who believed that, if they could get a common technology out, they could all compete using this common technology. The idea was not substantially different from that of an SDO, except that the consortium was focused on (usually) a single elaboration of a technology and its application to a particular market problem. The benefit of the consortium over the SDO was not, as many suppose, in the laxity of its rules (4). Rather, the benefit of the consortia was that all the players were like minded, and that the quest for consensus would not take as long. In many cases, the technology that was under consideration by a consortium was based on the specification of one of the participants and the intent of the other consortium participants was not so much to "correct the specification" as it was to get a product based upon the specification out into the market where it could be sold.

One of the key differentiators between the hardware and software business is that the hardware business is usually more stable than the software business. A computer mouse is a mouse and a keyboard is a keyboard. Within the software industry, however, the intended use of an Application Portability Interface (API) is nebulous. The API may be used to control whatever the mind of the software writer can conceive. At the same time, the market at which the software standardizer is aiming is not the end user (who usually doesn't care about software APIs) but is rather the application developer, who writes the application programs that the end users buy. This introduction of a two-tier demand system for standardization is also important in considering consortia since a single consortium might be appealing to a very select class of software writers. This is allowed in the consortium rules, while an SDO does not have this luxury.

The "two-tiered user" structure is important. The IT industry writes standards not for the end user, but for the intermediate layer of application providers and manufacturers. However, the attributes promised by the standards (commonality of information, the ability to intercommunicate) are what the end user (or the ultimate consumer) wants. The mechanisms to achieve these goals are provided at the discretion of the intermediate layer of providers. The IT industry—with its multiple layers of suppliers and differentiated users—is no different from many other industries. The complicating factor is, however, that the use and knowledge of IT is ubiquitous. At some point, nearly everyone has fancied herself or himself a programmer and used a personal computer to solve a problem or to lose information. Because of this ubiquity, the standards proposed for the industry to drive itself are part of the active consciousness of the end users, who may not understand them, but who do, at times, demand them.

And it is because of this ability to target a specific need among like-minded entities that consortia have their appeal. The consortia can create specifications in isolation from other standardization efforts and not be under pressure except from their sponsors to include other work in terms that may be deemed necessary by the larger market. While consortium sponsors used to claim that consortia can act more quickly than SDOs, this is not the real justification. What the consortium provides is a formalized structure (mimicking the SDOs) to which organizations can send people to accomplish a set task within a defined time frame. The key differentiators are the set task and defined time frame (5).

Within the SDO, there is no mechanism that mandates a schedule. All task forces or working groups or subcommittees file an expected calendar when they begin—and all routinely violate their schedules. While this may occur within a consortium it is usually less likely.

The players who participate know one another because they have the common interest (or else they wouldn't have paid the entrance fees to get in). The various management structures can be brought to bear on recalcitrant individuals or on recalcitrant organizations to meet schedule and to meet consensus. While the same bickering that occurs in SDOs may occur in a consortium, there is almost always a quick end to the problem because the players have a vested interest in success at multiple levels in the corporation.

This situation must be contrasted to that of the SDO, where there is not the level of commitment by participants to create on time. There is no guarantee of the depth of ''common purpose''; rather, there is only a guarantee of ''interested participants.'' The SDOs, by opening their doors to everyone, guarantee that they will be somewhat inefficient. In trade for this inefficiency, however, they are supposed to produce a superior product. Experience has proven that this is not the case. The marginal benefit of permitting anyone who has the money to participate (since, as noted previously, participation in an SDO does cost money) does not outweigh the benefit of schedule and fixity of purpose.

In a meeting of a consortium, I managed to gather experts from six of the world's leading database companies for a discussion of a possible standard. I designed a process that we would use, received a quick approval for the process, defined the technology, and then set to work. It took us approximately three months to come to a business-based decision on inclusion of a technology. We conducted only one vote (the final one) and held two meetings. If I had to reach the same decision within an SDO, I would have had to advertise the meetings to all interested parties, run the meetings with a process that was originally created for a different purpose, and then make sure that all opinions were considered, and so on and so on. I believe that the final decision would have been the same. The difference is that, under the rules of the consortia, I was allowed to gather subject-matter experts and come to an expert decision with them. The marginal benefits that would have accrued from having another constituency represented would not have substantially changed the decision of the group, nor made the technology better.

Another function that consortia have is the ability to create and run tests for implementations of their specifications. The ability to create test suites from the specifications—and to enforce adherence to the specifications by anyone who wants to claim conformance—is a powerful tool in the consortia arsenal. Because many consortia have the contractual ability to compel their participants to use the consortia specification in preference to another specification, the consortium specification can be adopted and implemented more quickly than an SDO standard, which usually comes out lacking these features. This ability to add the test and conformance mechanisms to their specification provides an added level of user safety. The guarantee of conformance to a specification—and specific redress if the conformance fails—is a powerful incentive to a manager who is seeking a guaranteed heterogeneous solution. Additionally, within the IT industry, this type of conformity guarantee is seen as the last step in solving the interoperability problem. The solution offered by many consortia is a "complete solution."

The idea of the complete solution is not as common in other industries as it is in the IT industries. The reason for the testing and verification of the implementation rests in the nature of the IT standards and specifications produced. In many cases, the specification or standard has options and the implementation of these options is left to the discretion of the provider. In many cases, the options are mutually exclusive options such as "connection-based service" or "connectionless service." If one provider selects a series of options (labeled A) and another provider selects a series of options (labeled B), both may be conforming but not interoperable, destroying the fundamental rationale for standardization (6). Conformance testing—to a specific profile—is the answer. However, test writing is an art, not a science, and the creation of tests is as expensive and time consuming as the creation of the standards. Even worse, the two disciplines are different. Standards creation requires one type of technologist, test creation requires a technologist with an entirely different mind set. So, when consortia offer a specification and the tests to validate implementation, they provide a higher level of service and meet more requirements than do SDOs.

In conclusion, the fundamental rationale for consortia is not, as many people have assumed, a laxness of rules and a possible speed of creation. Rather, it is a common fixity of purpose, the ability to make and adhere to a schedule, and the delivery of a complete specification (from business rationale to testing and conformance guarantee). Basically, the consortium is seen as a full-service organization, which provides better return on invested resources than does an SDO.

ALLIANCES: THE RATIONALE

Consortia came into their own in the late 1980s, and, within the IT industry, there were several significant ventures. COS was started by the providers of OSI standards and products to provide interoperability testing. This was followed by the Manufacturing Automation Protocol (MAP), driven by General Motors to create a profile for the Local Area Network (LAN), and Technical Office Profile (TOP) driven by Boeing to create an office profile. Other consortia began to spring up, driven by the promises of success that COS, MAP and TOP were delivering (7). These promises were followed by other promises, and by continuing growth in the consortia industry. At one time in the 1990 time frame, consortia were appearing at the rate of two per month, with a major consortia (with fees of $50,000 or more per year) appearing once a quarter. However, the progress that many of the consortia made was not enough to satisfy many of the participants in the IT provider arena.

This led to the most recent phenomena—alliances, which are seen as places in which vendors can collaborate on technology to provide a single interface specification used on multiple products. While alliances had existed before, the action that marketed the serious start of alliances was the creation of the Common Open Systems Environment (COSE), which was initiated in March of 1993. In the COSE announcement, the major systems firms in the UNIX arena announced that they would cooperate in the creation of a common desktop Graphical User Interface (GUI) and a common UNIX-based operating system, as well as starting activity to bring other technical activities under the same roof. The Common Desktop Environment (CDE) and the Unified UNIX operating system that emerged from this alliance were given to the X/Open, Ltd. (now The Open Group) consortia to both manage and create the necessary test suites.

The rationale for the alliance is that it can create a core specification very quickly. The Unified UNIX specification was created very quickly (but was tied up in external legal wrangling for nearly a year). The alliance, not being a long-lived body, tends to seek a place to position its work and will nearly always select a consortium, since the members of an alliance tend to favor consortia. The bias by alliances toward consortia over the SDO lies in the consortium's ability to provide those things that make the specification useful, such as tests, branding, and marketing. If the consortium is successful in convincing a large part of the market to accept the work of the alliance and the consortium, then there is a final step called the Publicly Available Specification (PAS), in which a publicly available and implemented specification is given to an SDO for acceptance as a formal de jure standard. The PAS goes through some of the hurdles of a formal standard. It is subject to widespread review and comment, but not— as a matter of practice—subject to change (8).

CONCLUSION

The alternatives to the formal process are becoming more numerous and, I believe, more viable as time passes. The tremendous amount of consensus that is currently being sought for IT standards—and generally most standards within the United States—is destroying their very effectiveness as viable tools to help guide and change the industry. As noted previously, the IT industry sees standards as a marketing agent—part of the product-attribute set. If these product attributes can be determined and codified more quickly elsewhere, the process that produces the formal standards will become more honored in the breach than in the observance. This will pertain to all standards in the industry over time, excepting those that deal with health or safety, or which are regulated and required by law.

There is a saying in business of "Lead, follow, or get out of the way." The current regime of standardization neither leads nor follows. It has, in many cases, become extraneous to real business. It is sought after only as an additive to an established process and, usually, product. If it is to become a serious part of the IT industry and, I believe, industry in general over the next ten years, it will have to take a page from the consortia and alliance programs and remember that the process that it selects serves the industry, and not the other way around. There are very few successful organizations in the commercial sector that have not undergone wrenching changes in process and organizational structure over the past ten years. The SDOs seem to believe that they are immune from this type of change. Just as companies who believed that they were immune to this change disappeared, so too will SDOs if they fail to change and understand that the informal processes work—and work well—for the industries that create them.

REFERENCES AND NOTES

1. See CF Cargill. A Modest Proposal for Business Based Standards. Computer Standards Evolution: Impact and Imperatives. Computer Standards Conference Proceedings 1988. Washington, DC: IEEE, 1988, pp. 60–4.
2. See, for example, HL Gabel, ed. Product Standardization and Competitive Strategy. Amsterdam: North Holland, 1987; CF, Cargill. Information Technology Standardization: Theory, Process, and Organizations. Bedford, MA: Digital Press, 1989; and RB Toth, ed. Handbook for Profits. New York: ANSI, 1990. Gabel's edition is a collection of pieces that has a heavy academic economic overtone, but nonetheless looks at standardization as an economic activity. Cargill's book is a combination of theory and experience in standardization in the IT industry. Toth's edition, published by ANSI, is a collection of pieces that provide an overview of how standardization fits into a company and how it can be leveraged for profitability.
3. See CF Cargill. Evolution and Revolution in Open Systems. StandardView 2(1): 1–10 March, 1995 for a more complete description of what happened in the IT industry—in

relation to major standardization movements. While there is much work done in describing the nature and state of the IT industry, few authors who write in this area have given credence to the problems and nature of standardization, which drove many of the successes (and failures) in the industry.

4. See A Updegrove. Consortia. Standard View 3(3): September 1995, for an analysis—from both a legal and procedural point of view—about the rigor of a consortium in enforcing rules and process.

5. An important distinction must be made here. A consortium is not necessarily a consortium. In the September 1992 Journal of the American Society for Information Science, Martin Weiss and Carl Cargill argue, in an article entitled "Consortia in the Standards Development Process," that there are three types of consortia. Of these three types (proof of concept, implementation, and application) the type that is being discussed is a hybrid "implementation" consortium. The three types correspond to the nature of the work being done. Proof of concept focuses on proving that a technical concept can be brought into existence. An implementation consortium focuses on creating a successful implementation from a known standard or specification, while the application consortium tends to focus on creating a specification to favor a specific group of companies. In the time since the article was published, application consortium have been largely overtaken by events, but some of their attributes—such as creation of specifications that serve the market—have been picked up by implementation consortium, who are looking at creating a specification that is capable of being implemented and having proof that the implementation corresponds to the specification.

6. Interoperability testing was not considered essential when the first major standardization efforts that required it began. It wasn't until the vendors had poured hundreds of millions of dollars of effort into OSI standardization that it became apparent that interoperability tests were needed. Since the late 1980s, nearly all IT standards have mandated that test points be inserted in standards, to facilitate the completion of tests.

7. In the larger scheme of things, and in relation to major currents in the IT industry, these three organizations delivered mainly hopes. Because they were user-driven technical consortia, they became detached from the mainstream of computing technology, which was driven by the providers. As Ken Olsen, the CEO of Digital once remarked about GM's MAP effort, "Digital doesn't build cars; GM shouldn't build computers."

8. The rationale for the unspoken rule restricting change to a PAS is very simple. If the PAS does represent the installed market base, a standard that was "sort of the same but different" would be largely useless, since the market would not change from current practice merely to be "standard." The rationale for the PAS is to bring specifications into standards, not to make the market change to embrace standards.

Appendix A: Standards Organizations and Acronyms

In the United States there are at least 750 organizations writing standards and probably as many or more in the rest of the world. The American domestic organizations are listed in a National Institute of Science and Technology (NIST) publication by Robert Toth, "Standards Activities of Organizations in the United States," NBS Special Publication 681, 1984. (A new edition bringing the material up to date is available). A work by Patricia L. Ricci and Linda Perry, "Standards—A Resource and Guide for Identification, Selection, and Acquisition," Stirtz, Bernards & Company, Minnesota, 1990 lists national, multinational, and international organizations.

We have selected from these sources a number of organizations that are frequently mentioned in standards literature and listed them alphabetically within each section by their acronyms or abbreviations for ready reference. The sections that follow are United States, Canadian, and Mexican organizations, followed by international and regional standards organizations.

U.S.-HEADQUARTERED ORGANIZATIONS AND ACRONYMS

AA   Aluminum Association, Washington, D.C. Serves the public as a source of noncommercial, industry-wide information on the aluminum industry.

AAMA   American Architectural Manufacturers Association, Des Plaines, IL. Produces technical standards for architectural aluminum products, doors, windows, etc.

AATCC   American Association of Textile Chemists and Colorists, Research Triangle Park, NC. Promotes increased knowledge of the application of dyes, chemicals, and auxiliaries in the textile industry.

A2LA   American Association for Laboratory Accreditation, Gaithersburg, MD. It is a multidisiplinary, private-sector laboratory accreditor and largest of its kind in the United States.

ABA   American Bankers Association, Washington, D.C. A national trade and professional association for 13,000 commercial banks.

ABYC   American Boat and Yacht Council, Millersville, MD. An independent, nonprofit membership organization representative of recreational boating including members from industry, insurance, government, and the boating public.

ACI   American Concrete Institute, Detroit, MI. Devotes its efforts to the solution of technical problems related to the design, construction, and maintenance of concrete and reinforced concrete structures.

AHAM   Association of Home Appliance Manufacturers, Chicago, IL. An international nonprofit association of companies manufacturing major and portable appliances.

AIA   Aerospace Industries Association of America, Washington, D.C. A national trade association representing U.S. companies engaged in research, development, and manufacture of such aerospace systems as aircraft, missiles, space craft, and space launch vehicles.

AIA   American Institute of Architects, Washington, D.C. Serves the needs and improves the capability of architects; conducts public relations and educational programs.

AIA   American Insurance Association, New York, NY. Represents 170 companies providing property and liability insurance and suretyship.

AIChE   American Institute of Chemical Engineers, New York, NY. Advances chemical engineering in theory and practice.

AISI   American Iron and Steel Institute, Washington, D.C. Activities include research, technology, engineering, collection and dissemination of statistics, and public distribution of information about the industry and its products.

ANS   American Nuclear Society, La Grange Park, IL. A nonprofit, international, scientific, engineering and education organization.

ANSI   American National Standards Institute, New York, NY, and Washington, DC. Approves and coordinates American National Standards from over 200 professional and technical soci-

	eties and trade associations that develop standards in the United States are members. There are more than 1,000 company members. ANSI is the official U.S. member body to the ISO and IEC international voluntary standards organizations.
API	American Petroleum Institute, Washington, D.C. A national association encompassing all branches of the petroleum industry and composed of approximately 300 corporate members and 6,500 individual members.
ARI	Air-Conditioning and Refrigeration Institute, Arlington, VA. Develops and promotes technical and international standards among its members in the United States.
ASCE	American Society of Civil Engineers, Washington, D.C. Develops standards on structural materials and performance for built systems.
ASHRAE	American Society of Heating, Refrigeration and Air Conditioning Engineers, Atlanta, GA. An international membership organization operated for the exclusive purpose of advancing the arts and sciences of heating, refrigeration, air conditioning, and ventilation.
ASME	American Society of Mechanical Engineers, New York, NY. An educational and scientific society organized to promote the art and science of mechanical engineering and the allied arts and sciences.
ASQ	American Society for Quality, (former ASQC, American Society for Quality Control) Milwaukee, WI. Advances the theory and practice of quality control including ANSI/ASQ 9000 quality standards.
ASSE	American Society of Safety Engineers, Des Plaines, IL. Composed of career-safety specialists organized to promote and foster the advancement of the profession and promotes several credential and certification programs.
ASTM	American Society for Testing and Materials, West Conshohocken, PA. A nonprofit organization formed to develop standards on characteristics and performance of materials, products, systems, and services; and to promote related knowledge. ASTM is one of the largest standards developers in the world. (N.B.: ASTM is not a testing organization or laboratory.)
AWWA	American Water Works Association, Denver, CO. Represents its 33,000 members in all areas concerning water supply.
BMA	Bicycle Manufacturers of America, Washington, D.C. A voluntary, private, nonprofit organization whose members produce 85% of all domestically manufactured bicycles.

BOCA   Building Officials and Code Administrators International, Coun-
            try Club Hills, IL. A nonprofit municipal-service membership
            organization involved in the field of code administration and
            enforcement, and community development.
CABO   Council of American Building Officials, Falls Church, VA. One
            of several nonprofit municipal service membership organiza-
            tions developing building codes and standards for municipal
            enforcement.
CBEMA   Computer and Business Equipment Manufacturers Association,
            see Information Technology Industry Council, ITIC as the suc-
            cessor organization.
COS   Corporation for Open Systems, McLean, VA. One of many non-
            traditional standards groups and developers of the open-sys-
            tems computer architecture and functional profile standards for
            computer systems.
CRI   Carpet and Rug Institute, Dalton, GA. A national trade association
            of 50 carpet and rug manufacturers and 125 companies supply-
            ing services and products to the industry. CRI has installation
            and environmental standards for carpet and rugs.
CSMA   Chemical Specialties Manufacturers Association, Washington,
            D.C. Four hundred companies involved in manufacturing, for-
            mulating, packaging, and marketing chemical–specialties prod-
            ucts.
EIA   Electronics Industries Association, Washington, D.C. A national
            trade association of companies involved in the manufacture of
            electronic components, equipment, and systems for consumer,
            industrial, and government applications.
FMRC   Factory Mutual Research Corporation, Norwood, MA. A wholly
            owned associate of the Factory Mutual System, providing loss-
            control standards development, basic research, applied re-
            search, and full-scale testing for industrial, commercial, and
            institutional policyholders.
HPMA   Hardwood Plywood Manufacturers Association, Reston, VA. A
            national trade association for hardwood and veneer manufac-
            turers and prefinishers of hardwood plywood.
ICBO   International Congress of Building Officials, Whittier, CA.
ICC   International Code Council, a coordinating group comprising sev-
            eral of the code organizations and whose aim is to harmonize
            the many diverse codes such as the electrical, plumbing, fire,
            and building codes.
ICSP   Interagency Committee on Standards Policy, secretariat is at
            NIST, Gaithersburg, MD. The federal government's forum that

brings together most agencies that have a role in standards development, usage, adoption into regulations, procurement standards, trade, and other related issues.

IEEE   Institute of Electrical and Electronic Engineers, New York, NY. A transactional professional society of 225,000 engineers and scientists in electrical engineering, electronics, and allied fields.

IES   Illuminating Engineering Society of North America, New York, NY. A technical society devoted to advance the art, science, and practice of illumination by investigation, evaluation, and dissemination of knowledge to consumers, producers, and general interest groups.

ISA   Instrument Society of America, Research Triangle Park, NC. Chartered as a nonprofit educational organization, its membership comprises more than 31,000 practitioners, scientists, educators, and students worldwide.

ITIC   Information Technology Industry Council (formerly CBEMA), Washington, D.C. A nonprofit organization that provides its members, computer and related-industry manufacturers, a forum for industry consultation and united action.

NAP   National Accreditation Panel, c/o RAB, ASQ, Milwaukee, WI. A joint operational panel of the American National Standards Institute and the Registrar Accreditation Board (ANSI/RAB), set up to manage their joint accreditation programs for quality management systems (QMS) and environmental management systems (EMS).

NEMA   National Electrical Manufacturers Association, Arlington, VA. A trade association of manufacturers of equipment and apparatus used for generation, transmission, distribution, and utilization of electric power; and developer of NEMA standards.

NESF   National Electrical Safety Foundation, Arlington, VA. Fosters electrical safety through public information and education programs, and operates independently but cooperatively with NEMA, US CPSC and other (see previous).

NFPA   National Fire Protection Association, Quincy, MA. An independent, nonprofit organization with approximately 67,000 members and 150 trade and professional association members worldwide. NFPA develops an extensive set of standards for fire-protection materials and systems.

NIBS   National Institute of Building Sciences, Washington, D.C. Congressionally authorized, nonprofit, nongovernmental scientific and technical institution that conducts research. Functions and responsibilities relate to building regulations.

NISO   National Information Standards Organization, Bethesda, MD. Developer of the ANSI Z39 series and other information standards of importance to librarians, publishers, and technical and information specialists.

NIST   National Institute of Standards and Technology, Gaithersburg, MD. A division of the U.S. Department of Commerce, known for promoting the development of standards for materials, weights and measures, standard reference materials (SRMs), and others.

NSF   National Sanitation Foundation, Ann Arbor, MI. A noncommercial, nonprofit standards, research, testing, and educational organization, with standards including those for water quality.

NSPI   National Spa and Pool Institute, Alexandria, VA. A national trade association of swimming pool builders, pool equipment manufacturers and suppliers, and others allied with the pool industry, also developer of pool and spa safety standards.

NSSN   A Global Standards Network on the Internet developed by ANSI. Its abbreviation derives from the former name National Standards Systems Network and can be found at its URL address http://www.nssn.org

NVCASE   National Voluntary Conformity Assessment System Evaluation, NIST, Gaithersburg, MD. The U.S. federal government's program for accrediting conformity-assessment systems, as part of our mutual-recognition agreement and equivalency negotiations with the European community and others.

OMB   Office of Management and Budget, Washington, D.C. OMB is author of circular A-119 prescribing government policy for "Federal Participation in the Development and Use of Voluntary Standards" by the agencies (see Appendix B).

OPEI   Outdoor Power Equipment Institute, Washington, D.C. A national trade association representing the domestic manufacturers of power lawn mowers, snow throwers, rototillers, garden tractors, and related equipment.

PPEMA   Portable Power Equipment Manufacturers Association, Bethesda, MD. A nonprofit industry organization representing manufacturers of gasoline and electrically powered chain saws and their component parts.

SBCCI   Southern Building Code Congress International, Birmingham, AL. Another of several building and life safety standards and code developers, used in municipal code enforcement.

SDOs   Standards developing organizations, an abbreviation used to refer to the several hundred private-sector standards developers active in the United States, examples being ASTM, NFPA, SAE, IEEE, ASME, etc. Contrast this with ANSI as the private-sector standards coordinator. ANSI is not an SDO.

SES   Standards Engineering Society, Miami, FL. A professional membership organization primarily of the United States and Canada, promoting the principles and practice of standardization, and professional standards education.

UL   Underwriters Laboratories, Northbrook, IL and other locations. A not-for-profit organization developing electrical, electrotechnical, and other safety-related standards. The UL mark and product or component listing is widely recognized and accepted by municipal enforcement and code officials.

USNC   United States National Committee for the International Electrotechnical Commission (IEC, see regional and international organizations). The USNC is administered through ANSI.

ORGANIZATIONS HEADQUARTERED IN CANADA

BNQ   Bureau de Normalisation du Quebec, Quebec. BNQ develops standards for use by Quebec and in Canada, and is one of the five accredited standards-writing organizations.

CGA   Canadian Gas Association, Don Mills, Ontario. CGA develops standards for use and safety of gas-related products and systems in Canada.

CGSB   Canadian General Standards Board, Ottawa, Ontario. CGSB develops many standards for building materials, products, and systems.

CSA   Canadian Standards Association, Rexdale, Ontario. CSA is the largest standards developer in Canada and also operates product listing and testing services for North America.

DND   Department of National Defence, Ottawa, Ontario. Developer of Canadian national defense standards.

SCC   Standards Council of Canada, Ottawa, Ontario. SCC is the government coordinator for the Canadian national standards system, official member body of ISO and IEC and accredits Canadian standards developers and testing laboratories.

ULC   Underwriters' Laboratories of Canada, Scarborough, Ontario. ULC is another of the accredited standards developers in Canada.

ORGANIZATIONS HEADQUARTERED IN MEXICO

ANCE   Mexican standards writing, testing, and certifying agency for the electrical sector.

CANACINTRA   A private-sector organization of business and industry, representing many of the various sector-specific chambers of commerce in Mexico.

DGN   Dirección General de Normas, Naucalpan de Juarez. The Mexican government's national standards body and official member to the ISO and IEC. Mexico's mandatory standards are known as NOMs or Official Mexican Standards. Voluntary standards are referred to as NMXs.

SECOFI   Mexican Department of Trade and Industry, or Ministry of Commerce and Industrial Development. In Spanish, Secretariá de Comercio y Fomento Industrial.

REGIONAL AND INTERNATIONAL STANDARDS ORGANIZATIONS

ALADI   Latin American Association for Integration (formerly ALAC, the Latin American Free Trade Association).

ANEC   European Association for the Coordination of Consumer Representation in Standards Work. A joint effort of both EC and EFTA, ANEC promotes the participation of consumers in European standardization work.

ARSO   African Regional Standards Organization, coordinating standards work among African nations.

ASEAN   Association of Southeast Asian Nations, comprising most of the nations in the Pacific Rim with key players being Australia and Japan.

CANENA   Council for Harmonization of Electrotechnical Standardization in North America, a tripartite organization among its United States, Canadian, and Mexican counterparts.

CASCO   The Conformity Assessment Committee of the ISO, Geneva, Switzerland, and responsible for ISO standards policy pertaining to certification, laboratory accreditation, quality management systems, and environmental management systems standards. See also COPOLCO.

CE   Certification mark for Europe, a European certification symbol for products complying with their directives and standards.

CEN   European Committee for Standardization, Brussels, Belgium. Developer of most European standards or ''norms'' (ENs) under contract to the European Union. European regional counterpart to the ISO. From the French, Comité Européen de Normalisation.

CENELEC   European Committee for Electrotechnical Standardization, Brussels, Belgium. Developer of the European regional electrotechnical standards under contract to the European Union. European regional counterpart to the IEC. Comité Européen de Normalisation Electrotechnique.

CI   Consumers International, The Hague, Netherlands. Coordinating body for national consumers unions and formerly known as International Organization of Consumers Unions (IOCU), of which Consumers Union is the U.S. member.

CIE   International Commission on Illumination, Vienna, Austria. Produces the CII international illumination and lighting standards.

CODEX   Codex Alimentarius, of the Food and Agriculture Organization (FAO), Rome, Italy. The Codex Alimentarius is a large body of international food and food additives safety and health standards.

COPANT   Pan American Standards Commission, Caracas, Venezuela. The organization promoting standards development and harmonization throughout the Americas, with the United States, Canada, Mexico, and many active Latin American and Caribbean nations involved in its work. The proper Spanish name is Comisión Panamerican de Normas Técnicas.

COPOLCO Consumer Policy Committee of the ISO, Geneva, Switzerland. One of four policy-development committees reporting to the ISO General Assembly, it promotes standards policy and technical work in many areas of consumer standards for safety, health, services, and quality. See also CASCO, the Conformity Assessment Committee of ISO.

ECOSA   European Consumer Safety Association, The Hague, Netherlands. Fosters European and some international activities to promote safety awareness, information, and education.

GATT   General Agreement on Tariffs and Trade, Geneva, Switzerland (see WTO or World Trade Organization, the successor to GATT).

EOQC   European Organization for Quality and Certification, Berne, Switzerland. Coordinator of quality standards, auditing, certification, and accreditation issues within Europe.

ETSI   European Telecommunications Standards Institute, Valbonne, France. Coordinator and developer of standards (ETSs) and related issues for telecommunications equipment, testing, and certification within Europe.

EU   European Union, Brussels, Belgium. Formerly known as the European Community (EC or EEC).

GCC   Gulf Cooperation Council, Riyadh, Saudi Arabia. A joint Arabian gulf-area standards cooperation and coordination council with its secretariat managed by SASO, the Saudi Arabian Standards Organization.

IEC   International Electrotechnical Commission, Geneva, Switzerland. Developer and publisher of about 2,000 to 2,500 standards and related documents in the fields of electrical components and electrical and electrotechnical products and systems. A companion organization to the ISO but in the electrotechnical area.

ILAC   International Laboratory Accreditation Conference. Fosters international collaboration on issues, guides, and standards related to laboratory accreditation.

ILO   International Labour Organization, Geneva, Switzerland. A UN-based organization with some interest in standards for labor and the workplace.

IMO   International Maritime Organization, London, United Kingdom. Publisher of standards for shipborne commerce and ship building.

ISO   International Organization for Standardization, Geneva, Switzerland. Developer and publisher of almost 10,000 voluntary international standards, the most well known being the ISO 9000 series for quality management systems (QMS) and 14000 series for environmental management systems (EMS). A companion organization to the IEC and ITU in all areas other than electrotechnical and telecommunications standards respectively.

ISONET   World Wide Network on standards, Geneva, Switzerland. Part of ISO, it helps coordinate and communicates with the national standards inquiry points within each ISO member body worldwide.

ITU   International Telecommunication Union, Geneva, Switzerland. The standards coordinator for international television, radio, telegraph, and most aspects of broadcast telecommunications standards, incorporating CCITT and CCIR. Now part of a tripartite coordination in voluntary international standards with the ISO and IEC.

NAFTA   North American Free Trade Agreement between Canada, Mexico, and the United States. Reducing trade barriers, harmonizing trade issues, and standards.

OECD   Organization for Economic Cooperation and Development, Paris, France. A treaty organization of those nations comprising the industrialized world (primarily Western Europe, the United States, Canada, and Japan), with several other countries holding ''observer'' status. OECD is active in standards research and study through its committees on consumer policy, trade, and others.

OIML   International Organization of Legal Metrology, Paris, France. A treaty organization fostering harmonized international measurement systems, primarily the International System of Units (SI) or the so-called "modernized metric system."

PASC   Pacific Area Standards Congress, a regional standards organization including the United States, Japan, Asian-Pacific countries, and others bordering the Pacific. ANSI is the U.S. member body to PASC.

QSAR   Quality System Assessment Recognition, ISO/IEC, Geneva, Switzerland. A voluntary international program of the ISO and IEC to encourage international recognition of ISO 9000 certificates worldwide.

UN-ECE   United Nations Economic Commission for Europe, Geneva, Switzerland. Studies and initiates regulations for automobile safety.

UNIDO   United Nations Industrial Development Organization, Vienna, Austria. UNIDO is a UN-based organization promoting, in part, standards training, national standards body development, and international standards among UN-member countries.

WIPO   World Intellectual Property Organization, Geneva, Switzerland. Registers and provides property-rights protection on intellectual property, design marks, and service marks, etc.

WTO   World Trade Organization (formerly General Agreement on Tariffs and Trade, GATT) Geneva, Switzerland. The WTO is a treaty organization and successor to the GATT. Most pertinent to standards professionals is the GATT and now WTO's Agreement on Technical Barriers to Trade (known as the "Standards Code" and the General Agreement on Trade in Services [GATS]).

Appendix B: OMB Circular A-119, Revised 1998

OMB Circular A-119; Federal Participation in the Development and Use of Voluntary Consensus Standards and in Conformity Assessment Activities

EXECUTIVE OFFICE OF THE PRESIDENT (EOP), Office of Management and Budget. ACTION: Final Revision of Circular A-119. Notice from the Federal Register, February 19, 1998.

SUMMARY

The Office of Management and Budget (OMB) revised Circular A-119 on federal use and development of voluntary standards. OMB revised this Circular in order to make the terminology of the Circular consistent with the National Technology Transfer and Advancement Act of 1995, to issue guidance to the agencies on making their reports to OMB, to direct the Secretary of Commerce to issue policy guidance for conformity assessment, and to make changes for clarity.

Direct any comments or inquiries to the Office of Information and Regulatory Affairs, Office of Management and Budget, NEOB Room 10236, Washington, D.C. 20503. Available at http://www.whitehouse.gov/WH/EOP/ omb or at (202) 395-7332. For further information contact Virginia Huth (202) 395-3785.

SUPPLEMENTARY INFORMATION:

I. Existing OMB Circular A-119

II. Authority

III. Notice and Request for Comments on Proposed Revision of OMB Circular 119-A

I. Existing OMB Circular A-119

Standards developed by voluntary consensus standards bodies are often appropriate for use in achieving federal policy objectives and in conducting federal activities, including procurement and regulation. The policies of OMB Circular A-119 are intended to: (1) Encourage federal agencies to benefit from the expertise of the private sector; (2) promote federal agency participation in such bodies to ensure creation of standards that are useable by federal agencies; and (3) reduce reliance on government-unique standards where an existing voluntary standard would suffice.

OMB Circular A-119 was last revised on October 20, 1993. This revision stated that the policy of the federal government, in its procurement and regulatory activities, is to: (1) ''rely on voluntary standards, both domestic and international, whenever feasible and consistent with law and regulation;'' (2) ''participate in voluntary standards bodies when such participation is in the public interest and is compatible with agencies' missions, authorities, priorities, and budget resources; '' and (3) ''coordinate agency participation in voluntary standards bodies so that the most effective use is made of agency resources and that the views expressed by such representatives are in the public interest and do not conflict with the interests and established views of the agencies.'' [See section 6 entitled 'Policy'].

II. Authority

Authority for this Circular is based on 31 USC 1111, which gives OMB broad authority to establish policies for the improved management of the Executive Branch. In February 1996, Section 12(d) of Public Law 104-113, the ''National Technology Transfer and Advancement Act of 1995,'' (or ''the Act'') was passed by the Congress in order to establish the policies of the existing OMB Circular A-119 in law. [See 142 Cong. Rec. H1264-1267 (daily ed. February 27, 1996) (statement of Rep. Morella); 142 Cong. Rec. S1078-1082 (daily ed. February 7, 1996) (statement of Sen. Rockefeller); 141 Cong. Rec. H14333-34 (daily ed. December 12, 1995) (statements of Reps. Brown and Morella)]. The purposes of Section 12(d) of the Act are: (1) To direct ''federal agencies to focus upon increasing their use of [voluntary consensus] standards whenever possible,'' thus, reducing federal procurement and operating costs; and (2) to authorize the National Institute of Standards and Technology (NIST) as the ''federal coordinator for government entities responsible for the development of technical standards and conformity assessment activities,'' thus eliminating "unnecessary duplication of conformity assessment activities.'' [See Cong. Rec. H1262 (daily ed. February 27, 1996) (statements of Rep. Morella)].

The Act gives the agencies discretion to use other standards in lieu of voluntary consensus standards where use of the latter would be "inconsistent with applicable law or otherwise impractical." However, in such cases, the head of an agency or department must send to OMB, through NIST, "an explanation of the reasons for using such standards." The Act states that beginning with fiscal year 1997, OMB will transmit to Congress and its committees an annual report summarizing all explanations received in the preceding year.

III. Notice and Request for Comments on Proposed Revision of OMB Circular A-119

On December 27, 1996, OMB published a "Notice and Request for Comments on Proposed Revision of OMB Circular A-119'' (61 FR 68312). The purpose of the proposed revision was to provide policy guidance to the agencies, to provide instructions on the new reporting requirements, to conform the Circular's terminology to the Act, and to improve the Circular's clarity and effectiveness. On February 10, 1997, OMB conducted a public meeting to receive comments and answer questions. In response to the proposed revision, OMB received comments from over 50 sources, including voluntary consensus standards bodies or standards development organizations (SDOs), industry organizations, private companies, federal agencies, and individuals. Although some commentators were critical of specific aspects of the proposed revision, the majority of commentators expressed support for the overall policies of the Circular and the approaches taken. The more substantive comments are available along with OMB's response. The Circular has also been converted into "plain English" format. Specifically, the following changes were made. We placed definitions where the term is first used; replaced the term *must* with *shall* where the intent was to establish a requirement; created a question and answer format using *you* and *I*; and added a Table of Contents. Accordingly, OMB Circular A-119 is revised as set forth below.

EXECUTIVE OFFICE OF THE PRESIDENT

Office of Management and Budget, Washington, D.C. 20503, February 10, 1998. Memorandum for Heads of Executive Departments and Agencies

THE CIRCULAR AND ITS CONTENTS

Federal Participation in the Development and Use of Voluntary Consensus Standards and in Conformity Assessment Activities

Revised OMB Circular A-119 establishes policies on Federal use and development of voluntary consensus standards and on conformity assessment activities. Public Law 104-113, the "National Technology Transfer and Advancement Act of 1995," codified existing policies in A-119, established reporting requirements, and authorized the National Institute of Standards and Technology to coordinate conformity assessment activities of the agencies. OMB is issuing this revision of the Circular in order to make the terminology of the Circular consistent with the National Technology Transfer and Advancement Act of 1995, to issue guidance to the agencies on making their reports to OMB, to direct the Secretary of Commerce to issue policy guidance for conformity assessment, and make changes for clarity.

BACKGROUND

1. What Is the Purpose of this Circular?

2. What Are the Goals of the Government Using Voluntary Consensus Standards?

DEFINITIONS OF STANDARDS

3. What Is a Standard?

4. What Are Voluntary Consensus Standards?

    a. Definition of voluntary, consensus standard.
       (1) Definition of voluntary, consensus standards body.
    b. Other types of standards.
       (1) Non-consensus standards, industry standards, company standards, or de facto standards.
       (2) Government-unique standards.
       (3) Standards mandated by law.

POLICY

5. Who Does this Policy Apply To?

6. What Is the Policy for Federal Use of Standards?

    a. When must my agency use voluntary consensus standards?
       (1) Definition of "Use."
       (2) Definition of "Impractical."

b. What must my agency do when such use is determined by my agency to be inconsistent with applicable law or otherwise impractical?

c. How does this policy affect my agency's regulatory authorities and responsibilities?

d. How does this policy affect my agency's procurement authority?

e. What are the goals of agency use of voluntary consensus standards?

f. What considerations should my agency make when it is considering using a standard?

g. Does this policy establish a preference between consensus and non-consensus standards that are developed in the private sector?

h. Does this policy establish a preference between domestic and international voluntary consensus standards?

i. Should my agency give preference to performance standards?

j. How should my agency reference voluntary consensus standards?

k. What if no voluntary consensus standard exists?

l. How may my agency identify voluntary consensus standards?

7. What Is the Policy for Federal Participation in Voluntary Consensus Standards Bodies?

a. What are the purposes of agency participation?

b. What are the general principles that apply to agency support?

c. What forms of support may my agency provide?

d. Must agency participants be authorized?

e. Does agency participation indicate endorsement of any decisions reached by voluntary consensus standards bodies?

f. Do agency representatives participate equally with other members?

g. Are there any limitations on participation by agency representatives?

h. Are there any limits on the number of federal participants in voluntary consensus standards bodies?

i. Is there anything else agency representatives should know?

j. What if a voluntary consensus standards body is likely to develop an acceptable, needed standard in a timely fashion?

8. What Is the Policy on Conformity Assessment?

MANAGEMENT AND REPORTING OF STANDARDS USE

9. What Is My Agency Required To Report?

10. How Does My Agency Manage and Report on Its Development and Use of Standards?

11. What Are the Procedures for Reporting My Agency's Use of Standards in Regulations?

12. What Are the Procedures for Reporting My Agency's Use of Standards in Procurements?

    a. How does my agency report the use of standards in procurements on a categorical basis?

    b. How does my agency report the use of standards in procurements on a transaction basis?

AGENCY RESPONSIBILITIES

13. What Are the Responsibilities of the Secretary of Commerce?

14. What Are the Responsibilities of the Heads of Agencies?

15. What Are the Responsibilities of Agency Standards Executives?

SUPPLEMENTARY INFORMATION

16. When Will this Circular Be Reviewed?

17. What Is the Legal Effect of this Circular?

18. Do You Have Further Questions?

BACKGROUND

1. What Is the Purpose of this Circular?

This Circular establishes policies to improve the internal management of the Executive Branch. Consistent with Section 12(d) of Pub. L. 104-113, the "National Technology Transfer and Advancement Act of 1995" (hereinafter "the Act"), this Circular directs agencies to use voluntary consensus standards in lieu of government-unique standards except where inconsistent with law or otherwise impractical. It also provides guidance for agencies participating in voluntary consensus standards bodies and describes procedures for satisfying the reporting requirements in the Act. The policies in this Circular are intended to reduce to a minimum the reliance by agencies on government-unique standards. These policies do not create the bases for discrimination in agency procurement or regulatory activities among standards developed in the private sector, whether or not they are developed by voluntary consensus standards bodies. Consistent with Section 12(b) of the Act, this Circular directs the Secretary of Commerce to issue guidance to the agencies in order to coordinate conformity assessment activities. This Circular replaces OMB Circular No. A-119 of October 20, 1993.

2. What Are the Goals of the Government in Using Voluntary Consensus Standards?

Many voluntary consensus standards are appropriate or adaptable for the Government's purposes. The use of such standards, whenever practicable and appropriate, is intended to achieve the following goals:

    a.   Eliminate the cost to the Government of developing its own standards and decrease the cost of goods procured and the burden of complying with agency regulation.

    b.   Provide incentives and opportunities to establish standards that serve national needs.

    c.   Encourage long-term growth for U.S. enterprises and promote efficiency and economic competition through harmonization of standards.

    d.   Further the policy of reliance upon the private sector to supply Government needs for goods and services.

DEFINITIONS OF STANDARDS

3. What Is a Standard?

    a.   The term standard, or technical standard as cited in the Act, includes all of the following:

        (1)   Common and repeated use of rules, conditions, guidelines or characteristics for products or related processes and production methods, and related management systems practices.

        (2)   The definition of terms; classification of components; delineation of procedures; specification of dimensions, materials, performance, designs, or operations; measurement of quality and quantity in describing materials, processes, products, systems, services, or practices; test methods and sampling procedures; or descriptions of fit and measurements of size or strength.

    b.   The term standard does not include the following:

        (1)   Professional standards of personal conduct.

        (2)   Institutional codes of ethics.

    c.   Performance standard is a standard as defined above that states requirements in terms of required results with criteria for verifying compliance but without stating the methods for achieving required results. A performance standard may define the functional requirements for the item, operational requirements, and/or interface and interchangeability characteristics. A performance standard may be viewed in juxtaposition to a prescriptive standard which may specify design requirements, such as materials to be used, how a requirement is to be achieved, or how an item is to be fabricated or constructed.

d. Non-government standard is a standard as defined above that is in the form of a standardization document developed by a private sector association, organization or technical society which plans, develops, establishes or coordinates standards, specifications, handbooks, or related documents.

4. What Are Voluntary Consensus Standards?

a. For purposes of this policy, voluntary consensus standards are standards developed or adopted by voluntary consensus standards bodies, both domestic and international. These standards include provisions requiring that owners of relevant intellectual property have agreed to make that intellectual property available on a non-discriminatory, royalty-free or reasonable royalty basis to all interested parties. For purposes of this Circular, ''technical standards that are developed or adopted by voluntary consensus standard bodies'' is an equivalent term.

(1) Voluntary consensus standards bodies are domestic or international organizations which plan, develop, establish, or coordinate voluntary consensus standards using agreed-upon procedures. For purposes of this Circular, ''voluntary, private sector, consensus standards bodies,'' as cited in Act, is an equivalent term. The Act and the Circular encourage the participation of federal representatives in these bodies to increase the likelihood that the standards they develop will meet both public and private sector needs. A voluntary consensus standards body is defined by the following attributes:

(i)   Openness.

(ii)   Balance of interest.

(iii) Due process.

(iv)   An appeals process.

(v)   Consensus, which is defined as general agreement, but not necessarily unanimity, and includes a process for attempting to resolve objections by interested parties, as long as all comments have been fairly considered, each objector

is advised of the disposition of his or her objection(s) and the reasons why, and the consensus body members are given an opportunity to change their votes after reviewing the comments.
b. Other types of standards, which are distinct from voluntary consensus standards, are the following:
    (1) "Non-consensus standards," "Industry standards," "Company standards," or "de facto standards," which are developed in the private sector but not in the full consensus process.
    (2) "Government-unique standards," which are developed by the government for its own uses.
    (3) Standards mandated by law, such as those contained in the United States Pharmacopeia and the National Formulary, as referenced in 21 U.S.C. 351.

POLICY

5. Who Does this Policy Apply To?

This Circular applies to all agencies and agency employees who use standards and participate in voluntary consensus standards activities, domestic and international, except for activities carried out pursuant to treaties. "Agency" means any executive department, independent commission, board, bureau, office, agency, Government-owned or controlled corporation or other establishment of the Federal Government. It also includes any regulatory commission or board, except for independent regulatory commissions insofar as they are subject to separate statutory requirements regarding the use of voluntary consensus standards. It does not include the legislative or judicial branches of the Federal Government.

6. What Is the Policy for Federal Use of Standards?

All federal agencies must use voluntary consensus standards in lieu of government-unique standards in their procurement and regulatory activities, except where inconsistent with law or otherwise impractical. In these circumstances, your agency must submit a report describing the reason(s) for its use of government-unique standards in lieu of voluntary consensus standards to the Office of Management and Budget (OMB) through the National Institute of Standards and Technology (NIST).

a. When must my agency use voluntary consensus standards?
    Your agency must use voluntary consensus standards, both domestic and international, in its regulatory and procurement activities in

lieu of government-unique standards, unless use of such standards would be inconsistent with applicable law or otherwise impractical. In all cases, your agency has the discretion to decline to use existing voluntary consensus standards if your agency determines that such standards are inconsistent with applicable law or otherwise impractical.

(1)   "Use" means incorporation of a standard in whole, in part, or by reference for procurement purposes, and the inclusion of a standard in whole, in part, or by reference in regulation(s).

(2)   "Impractical" includes circumstances in which such use would fail to serve the agency's program needs; would be infeasible; would be inadequate, ineffectual, inefficient, or inconsistent with agency mission; or would impose more burdens, or would be less useful, than the use of another standard.

b.   What must my agency do when such use is determined by my agency to be inconsistent with applicable law or otherwise impractical?
The head of your agency must transmit to the Office of Management and Budget (OMB), through the National Institute of Standards and Technology (NIST), an explanation of the reason(s) for using government-unique standards in lieu of voluntary consensus standards. For more information on reporting, see section 9.

c.   How does this policy affect my agency's regulatory authorities and responsibilities?
This policy does not preempt or restrict agencies' authorities and responsibilities to make regulatory decisions authorized by statute. Such regulatory authorities and responsibilities include determining the level of acceptable risk; setting the level of protection; and balancing risk, cost, and availability of technology in establishing regulatory standards. However, to determine whether established regulatory limits or targets have been met, agencies should use voluntary consensus standards for test methods, sampling procedures, or protocols.

d.   How does this policy affect my agency's procurement authority?
This policy does not preempt or restrict agencies' authorities and responsibilities to identify the capabilities that they need to obtain through procurements. Rather, this policy limits an agency's authority to pursue an identified capability through reliance on a government-unique standard when a voluntary consensus standard exists (see Section 6a).

e.   What are the goals of agency use of voluntary consensus standards?
Agencies should recognize the positive contribution of standards development and related activities. When properly conducted, standards development can increase productivity and efficiency in Government and

industry, expand opportunities for international trade, conserve resources, improve health and safety, and protect the environment.

f. What considerations should my agency make when it is considering using a standard?

When considering using a standard, your agency should take full account of the effect of using the standard on the economy, and of applicable federal laws and policies, including laws and regulations relating to antitrust, national security, small business, product safety, environment, metrication, technology development, and conflicts of interest. Your agency should also recognize that use of standards, if improperly conducted, can suppress free and fair competition; impede innovation and technical progress; exclude safer or less expensive products; or otherwise adversely affect trade, commerce, health, or safety. If your agency is proposing to incorporate a standard into a proposed or final rulemaking, your agency must comply with the ''Principles of Regulation'' (enumerated in Section 1(b)) and with the other analytical requirements of Executive Order 12866, ''Regulatory Planning and Review.''

g. Does this policy establish a preference between consensus and non-consensus standards that are developed in the private sector?

This policy does not establish a preference among standards developed in the private sector. Specifically, agencies that promulgate regulations referencing non-consensus standards developed in the private sector are not required to report on these actions, and agencies that procure products or services based on non-consensus standards are not required to report on such procurements. For example, this policy allows agencies to select a non-consensus standard developed in the private sector as a means of establishing testing methods in a regulation and to choose among commercial-off-the-shelf products, regardless of whether the underlying standards are developed by voluntary consensus standards bodies or not.

h. Does this policy establish a preference between domestic and international voluntary consensus standards?

This policy does not establish a preference between domestic and international voluntary consensus standards. However, in the interests of promoting trade and implementing the provisions of international treaty agreements, your agency should consider international standards in procurement and regulatory applications.

i. Should my agency give preference to performance standards?

In using voluntary consensus standards, your agency should give preference to performance standards when such standards may reasonably be used in lieu of prescriptive standards.

j. How should my agency reference voluntary consensus standards?
    Your agency should reference voluntary consensus standards, along
    with sources of availability, in appropriate publications, regulatory or-
    ders, and related internal documents. In regulations, the reference must
    include the date of issuance. For all other uses, your agency must deter-
    mine the most appropriate form of reference, which may exclude the
    date of issuance as long as users are elsewhere directed to the latest
    issue. If a voluntary standard is used and published in an agency docu-
    ment, your agency must observe and protect the rights of the copyright
    holder and any other similar obligations.

k. What if no voluntary consensus standard exists?
    In cases where no voluntary consensus standards exist, an agency may
    use government-unique standards (in addition to other standards, see
    Section 6g) and is not required to file a report on its use of government-
    unique standards. As explained above (see Section 6a), an agency may
    use government-unique standards in lieu of voluntary consensus stan-
    dards if the use of such standards would be inconsistent with applicable
    law or otherwise impractical; in such cases, the agency must file a report
    under Section 9a regarding its use of government-unique standards.

l. How may my agency identify voluntary consensus standards?
    Your agency may identify voluntary consensus standards through data-
    bases of standards maintained by the National Institute of Standards
    and Technology (NIST), or by other organizations including voluntary
    consensus standards bodies, other federal agencies, or standards pub-
    lishing companies.

7. What Is the Policy for Federal Participation in Voluntary Consensus Standards Bodies?

Agencies must consult with voluntary consensus standards bodies, both domestic
and international, and must participate with such bodies in the development of
voluntary consensus standards when consultation and participation is in the public
interest and is compatible with their missions, authorities, priorities, and budget
resources.

a. What are the purposes of agency participation?
    Agency representatives should participate in voluntary consensus stan-
    dards activities in order to accomplish the following purposes:
    (1) Eliminate the necessity for development or maintenance of sepa-
         rate Government-unique standards.
    (2) Further such national goals and objectives as increased use of the
         metric system of measurement; use of environmentally sound and
         energy efficient materials, products, systems, services, or prac-
         tices; and improvement of public health and safety.

b. What are the general principles that apply to agency support? Agency support provided to a voluntary consensus standards activity must be limited to that which clearly furthers agency and departmental missions, authorities, priorities, and is consistent with budget resources. Agency support must not be contingent upon the outcome of the standards activity. Normally, the total amount of federal support should be no greater than that of other participants in that activity, except when it is in the direct and predominant interest of the Government to develop or revise a standard, and its timely development or revision appears unlikely in the absence of such support.

c. What forms of support may my agency provide? The form of agency support, may include the following:

(1) Direct financial support; e.g., grants, memberships, and contracts.

(2) Administrative support; e.g., travel costs, hosting of meetings, and secretarial functions.

(3) Technical support; e.g., cooperative testing for standards evaluation and participation of agency personnel in the activities of voluntary consensus standards bodies.

(4) Joint planning with voluntary consensus standards bodies to promote the identification and development of needed standards.

(5) Participation of agency personnel.

d. Must agency participants be authorized? Agency employees who, at Government expense, participate in standards activities of voluntary consensus standards bodies on behalf of the agency must do so as specifically authorized agency representatives. Agency support for, and participation by agency personnel in, voluntary consensus standards bodies must be in compliance with applicable laws and regulations. For example, agency support is subject to legal and budgetary authority and availability of funds. Similarly, participation by agency employees (whether or not on behalf of the agency) in the activities of voluntary consensus standards bodies is subject to the laws and regulations that apply to participation by federal employees in the activities of outside organizations. While we anticipate that participation in a committee that is developing a standard would generally not raise significant issues, participation as an officer, director, or trustee of an organization would raise more significant issues. An agency should involve its agency ethics officer, as appropriate, before authorizing support for or participation in a voluntary consensus standards body.

e. Does agency participation indicate endorsement of any decisions reached by voluntary consensus standards bodies? Agency participation in voluntary consensus standards bodies does not necessarily connote agency agreement with, or endorsement of, decisions reached by such organizations.

f. Do agency representatives participate equally with other members? Agency representatives serving as members of voluntary consensus standards bodies should participate actively and on an equal basis with other members, consistent with the procedures of those bodies, particularly in matters such as establishing priorities, developing procedures for preparing, reviewing, and approving standards, and developing or adopting new standards. Active participation includes full involvement in discussions and technical debates, registering of opinions and, if selected, serving as chairpersons or in other official capacities. Agency representatives may vote, in accordance with the procedures of the voluntary consensus standards body, at each stage of the standards development process unless prohibited from doing so by law or their agencies.

g. Are there any limitations on participation by agency representatives? In order to maintain the independence of voluntary consensus standards bodies, agency representatives must refrain from involvement in the internal management of such organizations (e.g., selection of salaried officers and employees, establishment of staff salaries, and administrative policies). Agency representatives must not dominate such bodies, and in any case are bound by voluntary consensus standards bodies' rules and procedures, including those regarding domination of proceedings by any individual. Regardless, such agency employees must avoid the practice or the appearance of undue influence relating to their agency representation and activities in voluntary consensus standards bodies.

h. Are there any limits on the number of federal participants in voluntary consensus standards bodies? The number of individual agency participants in a given voluntary standards activity should be kept to the minimum required for effective representation of the various program, technical, or other concerns of federal agencies.

i. Is there anything else agency representatives should know? This Circular does not provide guidance concerning the internal operating procedures that may be applicable to voluntary consensus standards bodies because of their relationships to agencies under this Circular. Agencies should, however, carefully consider what laws or rules may apply in a particular instance because of these relationships. For example, these relationships may involve the Federal Advisory Committee Act, as amended (5 U.S.C. App. I), or a provision of an authorizing statute for a particular agency.

j. What if a voluntary consensus standards body is likely to develop an acceptable, needed standard in a timely fashion?

If a voluntary consensus standards body is in the process of developing or adopting a voluntary consensus standard that would likely be lawful and practical for an agency to use, and would likely be developed or adopted on a timely basis, an agency should not be developing its own government-unique standard and instead should be participating in the activities of the voluntary consensus standards body.

8. What Is the Policy on Conformity Assessment?

Section 12(b) of the Act requires NIST to coordinate Federal, State, and local standards activities and conformity assessment activities with private sector standards activities and conformity assessment activities, with the goal of eliminating unnecessary duplication and complexity in the development and promulgation of conformity assessment requirements and measures. To ensure effective coordination, the Secretary of Commerce must issue guidance to the agencies.

MANAGEMENT AND REPORTING OF STANDARDS USE

9. What Is My Agency Required to Report?

a. As required by the Act, your agency must report to NIST, no later than December 31 of each year, the decisions by your agency in the previous fiscal year to use government-unique standards in lieu of voluntary consensus standards. If no voluntary consensus standard exists, your agency does not need to report its use of government-unique standards. (In addition, an agency is not required to report on its use of other standards. See Section 6g.) Your agency must include an explanation of the reason(s) why use of such voluntary consensus standard would be inconsistent with applicable law or otherwise impractical, as described in Sections 11b(2), 12a(3), and 12b(2) of this Circular. Your agency must report in accordance with format instructions issued by NIST.

b. Your agency must report to NIST, no later than December 31 of each year, information on the nature and extent of agency participation in the development and use of voluntary consensus standards from the previous fiscal year. Your agency must report in accordance with format instructions issued by NIST. Such reporting must include the following:

(1) The number of voluntary consensus standards bodies in which there is agency participation, as well as the number of agency employees participating.

(2) The number of voluntary consensus standards the agency has used since the last report, based on the procedures set forth in sections 11 and 12 of this Circular.

(3) Identification of voluntary consensus standards that have been substituted for government-unique standards as a result of an agency review under section 15b(7) of this Circular.

(4) An evaluation of the effectiveness of this policy and recommendations for any changes.

c. No later than the following January 31, NIST must transmit to OMB a summary report of the information received.

10. How Does My Agency Manage and Report On Its Development and Use of Standards?

Your agency must establish a process to identify, manage, and review your agency's development and use of standards. At minimum, your agency must have the ability to (1) report to OMB through NIST on the agency's use of government-unique standards in lieu of voluntary consensus standards, along with an explanation of the reasons for such non-usage, as described in section 9a, and (2) report on your agency's participation in the development and use of voluntary consensus standards, as described in section 9b. This policy establishes two ways, category based reporting and transaction based reporting, for agencies to manage and report their use of standards. Your agency must report all uses of standards in one or both ways.

11. What Are the Procedures for Reporting My Agency's Use of Standards in Regulations?

Your agency should use transaction based reporting if your agency issues regulations that use or reference standards. If your agency is issuing or revising a regulation that contains a standard, your agency must follow these procedures:

a. Publish a request for comment within the preamble of a Notice of Proposed Rulemaking (NPRM) or Interim Final Rule (IFR). Such request must provide the appropriate information, as follows:

(1) When your agency is proposing to use a voluntary consensus standard, provide a statement which identifies such standard.

(2) When your agency is proposing to use a government-unique standard in lieu of a voluntary consensus standard, provide a statement which identifies such standards and provides a preliminary explanation for the proposed use of a government-unique standard in lieu of a voluntary consensus standard.

(3) When your agency is proposing to use a government-unique standard, and no voluntary consensus standard has been identified, a statement to that effect and an invitation to identify any such standard and to explain why such standard should be used.

b. Publish a discussion in the preamble of a Final Rulemaking that restates the statement in the NPRM or IFR, acknowledges and summarizes any comments received and responds to them, and explains the agency's final decision. This discussion must provide the appropriate information, as follows:

(1) When a voluntary consensus standard is being used, provide a statement that identifies such standard and any alternative voluntary consensus standards which have been identified.

(2) When a government-unique standard is being used in lieu of a voluntary consensus standard, provide a statement that identifies the standards and explains why using the voluntary consensus standard would be inconsistent with applicable law or otherwise impractical. Such explanation must be transmitted in accordance with the requirements of Section 9a.

(3) When a government-unique standard is being used, and no voluntary consensus standard has been identified, provide a statement to that effect.

12. What Are the Procedures for Reporting My Agency's Use of Standards in Procurements?

To identify, manage, and review the standards used in your agency's procurements, your agency must either report on a categorical basis or on a transaction basis.

a. How does my agency report the use of standards in procurements on a categorical basis?

Your agency must report on a category basis when your agency identifies, manages, and reviews the use of standards by group or category. Category based reporting is especially useful when your agency either conducts large procurements or large numbers of procurements using government-unique standards, or is involved in long-term procurement contracts which require replacement parts based on government-unique standards. To report use of government-unique standards on a categorical basis, your agency must:

(1) Maintain a centralized standards management system that identifies how your agency uses both government-unique and voluntary consensus standards.

(2) Systematically review your agency's use of government-unique standards for conversion to voluntary consensus standards.

(3) Maintain records on the groups or categories in which your agency uses government-unique standards in lieu of voluntary

consensus standards, including an explanation of the reasons for such use, which must be transmitted according to Section 9a.

    (4)   Enable potential offerors to suggest voluntary consensus standards that can replace government-unique standards.

b. How does my agency report the use of standards in procurements on a transaction basis?

Your agency should report on a transaction basis when your agency identifies, manages, and reviews the use of standards on a transaction basis rather than a category basis. Transaction based reporting is especially useful when your agency conducts procurement mostly through commercial products and services, but is occasionally involved in a procurement involving government-unique standards. To report use of government-unique standards on a transaction basis, your agency must follow the following procedures:

    (1)   Each solicitation which references government-unique standards must:

        (i)   Identify such standards.

        (ii) Provide potential offerors an opportunity to suggest alternative voluntary consensus standards that meet the agency's requirements.

    (2)   If such suggestions are made and the agency decides to use government-unique standards in lieu of voluntary consensus standards, the agency must explain in its report to OMB as described in Section 9a why using such voluntary consensus standards is inconsistent with applicable law or otherwise impractical.

c. For those solicitations that are for commercial-off-the-shelf products (COTS), or for products or services that rely on voluntary consensus standards or non-consensus standards developed in the private sector, or for products that otherwise do not rely on government-unique standards, the requirements in this section do not apply.

AGENCY RESPONSIBILITIES

13. What Are the Responsibilities of the Secretary of Commerce?

The Secretary of Commerce:

    a.   Coordinates and fosters executive branch implementation of this Circular and, as appropriate, provides administrative guidance to assist agencies in implementing this Circular including guidance on identifying voluntary consensus standards bodies and voluntary consensus standards.

b.   Sponsors and supports the Interagency Committee on Standards Policy (ICSP), chaired by the National Institute of Standards and Technology, which considers agency views and advises the Secretary and agency heads on the Circular.

c.   Reports to the Director of OMB concerning the implementation of the policy provisions of this Circular.

d.   Establishes procedures for agencies to use when developing directories described in Section 15b(5) and establish procedures to make these directories available to the public.

e.   Issues guidance to the agencies to improve coordination on conformity assessment in accordance with section 8.

14. What Are the Responsibilities of the Heads of Agencies?

The Heads of Agencies:

a.   Implement the policies of this Circular in accordance with procedures described.

b.   Ensure agency compliance with the policies of the Circular.

c.   In the case of an agency with significant interest in the use of standards, designate a senior level official as the Standards Executive who will be responsible for the agency's implementation of this Circular and who will represent the agency on the ICSP.

d.   Transmit the annual report prepared by the Agency Standards Executive as described in Sections 9 and 15b(6).

15. What Are the Responsibilities of Agency Standards Executives?

An Agency Standards Executive:

a.   Promotes the following goals:

(1)   Effective use of agency resources and participation.

(2)   The development of agency positions that are in the public interest and that do not conflict with each other.

(3)   The development of agency positions that are consistent with administration policy.

(4)   The development of agency technical and policy positions that are clearly defined and known in advance to all federal participants on a given committee.

b.   Coordinates his or her agency's participation in voluntary consensus standards bodies by:

(1) Establishing procedures to ensure that agency representatives who participate in voluntary consensus standards bodies will, to the extent possible, ascertain the views of the agency on matters of paramount interest and will, at a minimum, express views that are not inconsistent or in conflict with established agency views.

(2) To the extent possible, ensuring that the agency's participation in voluntary consensus standards bodies is consistent with agency missions, authorities, priorities, and budget resources.

(3) Ensuring, when two or more agencies participate in a given voluntary consensus standards activity, that they coordinate their views on matters of paramount importance so as to present, whenever feasible, a single, unified position and, where not feasible, a mutual recognition of differences.

(4) Cooperating with the Secretary in carrying out their responsibilities under this Circular.

(5) Consulting with the Secretary, as necessary, in the development and issuance of internal agency procedures and guidance implementing this Circular, including the development and implementation of an agency-wide directory identifying agency employees participating in voluntary consensus standards bodies and the identification of voluntary consensus standards bodies.

(6) Preparing, as described in Section 9, a report on uses of government-unique standards in lieu of voluntary consensus standards and a report on the status of agency standards policy activities.

(7) Establishing a process for ongoing review of the agency's use of standards for purposes of updating such use.

(8) Coordinating with appropriate agency offices (e.g., budget and legal offices) to ensure that effective processes exist for the review of proposed agency support for, and participation in, voluntary consensus standards bodies, so that agency support and participation will comply with applicable laws and regulations.

SUPPLEMENTARY INFORMATION

16. When Will this Circular Be Reviewed?

This Circular will be reviewed for effectiveness by the OMB three years from the date of issuance.

17. What Is the Legal Effect of this Circular?

Authority for this Circular is based on 31 USC 1111, which gives OMB broad authority to establish policies for the improved management of the Executive Branch. This Circular is intended to implement Section 12(d) of Public Law 104-113 and to establish policies that will improve the internal management of the Executive Branch. This Circular is not intended to create delay in the administrative process, provide new grounds for judicial review, or create new rights or benefits, substantive or procedural, enforceable at law or equity by a party against the United States, its agencies or instrumentalities, or its officers or employees.

18. Do You Have Further Questions?

For information concerning this Circular, contact the Office of Management and Budget, Office of Information and Regulatory Affairs: Telephone 202/395-3785.

REFERENCE

Federal Register Notice: February 19, 1998 (Volume 63, Number 33), Pages 8545–8558. From the Federal Register Online via GPO Access [wais.access.gpo.gov] [DOCID: fr19fe98–146]

Index

301